D1544215

ZOOGEOMORPHOLOGY

Animals as geomorphic agents have primarily been considered "curiosities" in the literature of geomorphology, whose spatial and quantitative influences have been seen as both limited and minor. *Zoogeomorphology: Animals as Geomorphic Agents* examines the distinct geomorphic influences of invertebrates, ectothermic vertebrates, birds, and mammals, and demonstrates the importance of animals as landscape sculptors. Specific processes associated with the diversity of animal influences in geomorphology are examined, including burrowing and denning, nesting, lithophagy and geophagy, wallowing and trampling, food caching, excavating for food, and dam building by beavers. Particular emphasis is placed on terrestrial animals, although aquatic animals are also discussed where appropriate.

This book, the only one available wholly devoted to this topic, will interest graduate students and professional research workers in geomorphology, ecology, environmental science, physical geography, and geology.

ZOOGEOMORPHOLOGY
ANIMALS AS GEOMORPHIC AGENTS

DAVID R. BUTLER

University of North Carolina, Chapel Hill

CAMBRIDGE
UNIVERSITY PRESS

To Janet and William Butler

Published by the Press Syndicate of the University of Cambridge
The Pitt Building, Trumpington Street, Cambridge CB2 1RP
40 West 20th Street, New York, NY 10011-4211, USA
10 Stamford Road, Oakleigh, Melbourne 3166, Australia

© Cambridge University Press 1995

First published 1995

Printed in the United States of America

Library of Congress Cataloging-in-Publication Data
Butler, David R.
    Zoogeomorphology : animals as geomorphic agents / David R. Butler
        p.   cm.
    Includes bibliographical references and index.
    ISBN 0-521-43343-6 (hc)
    1. Biogeomorphology. I. Title.
    QH542.5.888   1995
    591.52′2 – dc20   94–44726
                                                            CIP

A catalog record for this book is available from the British Library

ISBN 0-521-43343-6 hardback

Contents

Acknowledgments

The creation of a book such as this is a long process that has its roots in many years of observation, classwork training, research, and travel. Many people and organizations have directly or indirectly aided in the development of my interests in animals as geomorphic agents, and deserve recognition. I would first like to express my gratitude to several individuals who served as mentors during my training in geomorphology: Dr. Nicholas Bariss and Dr. Jack Shroder, University of Nebraska–Omaha; and Dr. Curt Sorenson, Dr. Bill Johnson, and Dr. Wake Dort, University of Kansas. I also thank Dr. Jack Vitek for his friendship and guidance during my early postgraduate years.

Field assistants and graduate students are some of the unsung heroes of academia. Often for no salary, a limited food budget, and marginal housing, they serve just to be there and to learn. I am very grateful to Susan Panciera, Jack and Lori Oelfke, Dan Brown, Katherine Schipke, Dave Cairns, and Bill Welsh for their help, comraderie, comments, and insight, and critical mass when hiking in grizzly terrain. I also thank Marilyn Wyrick for her expert library ferreting.

Financial assistance for my work on zoogeomorphic topics has come from a variety of sources. I thank Dr. Jim Fisher and the Department of Geography, University of Georgia, for their support over the six-year period of my residence there. The Department of Geography at the University of North Carolina–Chapel Hill, and Dr. John Florin, Chair, have also provided solid support for this project, not the least of which was in the form of a Semester of Research and Study during the fall of 1993. Several grants and agencies funded fieldwork in Glacier National Park, Montana, which exposed me to a diversity of geomorphic effects produced by animals. I thank the University of Georgia Research Foundation, Inc. (summer 1987), the National Geographic Society (Grant 3831-88; summer 1988), Dr. Steve

Walsh and his NASA grant (summer 1990), the Geography and Regional Science Division of the National Science Foundation (Grant No. SES-9109837) (summers 1991–2), the J. D. Eyre Fund of the Department of Geography, UNC–Chapel Hill (January and October 1993), and the UNC University Research Council (October 1993).

Officials in the Science Office of Glacier National Park provided research collecting permits, logistical support, and housing. I particularly thank Clifford Martinka, former Chief Scientist and current Senior Scientist, for his efforts during my early research forays into the park; Carl Key, Park Geographer; and Dr. Dan Fagri, director of the park's climate change program. Ms. Beth Ludeau-Denevan of the Park Library has also been most helpful.

Local individuals in the Glacier Park area have provided valuable information and advice about field sites and conditions, accommodations, and access through the years. Of particular note were Suzie, Sandy, and Megan during the summer of 1988, who kept the cold ones coming and provided occasional additional ones at no cost!

Dr. Robin Smith, Life Sciences Editor, and Michael Gnat, Production Editor, have been extremely helpful and supportive of my efforts. I sincerely thank them both.

Dr. Steve Walsh and Dr. George Malanson have, over the past twelve years, been constant sources of camaraderie, friendship, and inspiration. Each has been "infected" by the Glacier Park virus, for which I claim full responsibility. They have collaborated together and separately with me on many of the field efforts that resulted in my observations of animals as geomorphic agents, and I thank them sincerely. They each deserve a jug of wine, a loaf of bread, and a huge hunk of Pecorino!

Finally, my family has, through the years, been my most consistent source of support, love, and inspiration. My parents, Ray and Marian Butler, instilled in me a love for the natural world, and have been supportive of my research efforts even when they had to defend to their friends why their son was studying beavers! My brother, Mike Butler, has also been there as a supporter and frequent field companion. Most important, my wife and son, Janet and William Butler, have provided love and unwavering support. They have understood why "Daddy has to go to Montana *again.*" No one could be luckier than I have been.

1
Introduction

Geomorphology is the study of surface processes and landforms (Easterbrook 1993). Most geomorphologists, including Easterbrook, consider that geomorphology also encompasses the evolution of landforms and interpretations as to their origin. Geomorphology therefore examines the processes currently or recently operative on the earth's surface that erode, transport, and deposit sediment and that create landforms. Rhoads and Thorn (1993, p. 288) succinctly summarized the discipline of geomorphology while extending its reach both temporally and spatially, by stating that the discipline is "the study of past, present, and future landforms, landform assemblages (physical landscapes), and surficial processes on the earth and other planets."

In typical introductory geomorphology textbooks, a variety of surficial and internal processes are described. A common list of topics covered in such books would include diastrophic forces of folding and faulting, internal and surface volcanism, weathering and soil development, gravity and mass movement, the work of running water on and under the surface, the work of glacial ice and ground ice, wind, and wave and current action. (See Gregory [1988] for a discussion of recent curriculum trends in geomorphology.) Unfortunately, these processes are frequently presented as if they were operative in a sterile, nonliving void. When the biosphere is even acknowledged or recognized in such works, it is usually restricted to large-scale overviews of geomorphic processes operative in different climatic regions or biomes (i.e., climatic geomorphology).

Animals are, however, a conspicuous element of the earth's physical landscape and its environmental systems, but one that is typically glossed over or completely ignored in earth-science texts and classes at the primary- and secondary-school levels. Even in college and university classes in physical geography, animals are at best mentioned parenthetically. Only in ad-

1

vanced university classes in biogeography are animals closely examined as important elements of the physical landscape, and their role as geomorphic elements is not typically emphasized in such classes.

Why has the role of animals been ignored in most twentieth-century geomorphology? Viles (1988a) identified the significance of Davisian geomorphology as the primary reason. Davisian models of landscape development focused in large part on landscape history at the regional macro scale of investigation, rather than landscape *processes* operative at the meso or micro scale. With the decline of Davisian geomorphology after World War II, the study of processes received more attention at the meso scale. Processes operative at the micro scale, such as biological organisms, have only recently begun to be examined closely (Viles 1988a).

Even with the increased emphasis on micro-scale processes, however, geomorphologists have not suddenly "jumped on the bandwagon" of studying the effects of animals on the landscape. This is probably a result of the perspective of many geomorphologists trained in the earth and physical sciences rather than the biological sciences. The background and training of geomorphologists, then, translates into an approach to the living landscape in which they typically examine how surface processes affect environmental and ecological systems, rather than how elements of the biological landscape affect and may act as geomorphic processes (e.g., see Swanson et al. 1988). A certain parochial attitude among some geologically trained geomorphologists may also be a factor, reflected by an unfamiliarity with, or unwillingness to utilize, literature sources from outside the earth sciences. The cross-disciplinary nature of the discipline of geography makes geomorphologists trained in that tradition perhaps more aware of, and attentive to, scientific literature in the biological sciences.

As the twentieth century draws to a close, commonly used geomorphology texts (Thornbury 1969; Twidale 1975; Rice 1977; Chorley, Schumm, and Sugden 1985; Selby 1985; Ritter 1986; Rohdenburg 1989; Bloom 1991; Summerfield 1991; Easterbrook 1993), as well as summaries of different national research efforts in geomorphology (Ahnert 1989; Le Groupe Français de Géomorphologie 1989; Suzuki 1989), still completely ignore the fundamental role of animals as geomorphic agents of erosion, transportation, and deposition (but see Lobeck [1939] for an example of an early geomorphology text that *did* include coverage of animals). The only exception to this rule is in the study of coral reefs as biogenic landforms, which occasionally gains mention in geomorphology textbooks and has a long tradition of study in geomorphology going back (again) to the later works of William Morris

Davis (Guilcher 1988). Indeed, the scholarly journal *Coral Reefs* is devoted entirely to various aspects of the study of these unusual landforms.

This is not to say that geomorphologists have been completely derelict in the study of animals as agents of erosion. Several papers have appeared since the mid-1970s in major journals of geomorphology and physical geography dealing with topics such as the geomorphic role of beavers, ground squirrels, rabbits, badgers, porcupines, grizzly bears (Figs. 1.1, 1.2), and some invertebrates. Several chapters in Viles's (1988d) compilation on biogeomorphology examined aspects of animals as erosional agents, but the emphasis there was primarily on geomorphological interactions with vegetation.(Also see the book edited by Thornes [1990], including Viles's [1990] paper therein.)

Interestingly, however, animal ecologists and wildlife managers have also recently made strides in the study of animals as geomorphic agents. These studies typically examine animals from a paradigmatic framework of landscape ecology or wildlife behavioral ecology, however, rather than from a truly geomorphological perspective.

My goals in writing this book are to bring together this diverse research literature covering the study of animals as geomorphic agents, whether from a process geomorphology or a landscape/wildlife ecology paradigmatic background, with my own experiences in the study of animals as geomorphic agents; to illustrate the levels of geomorphic work carried on in the animal kingdom; and to compare those levels with the more traditionally studied geomorphic processes. It has been suggested that "while much of biogeomorphology is obviously fun, it is not so obviously fundamental" (Cox 1989, p. 623). On the other hand, Johnson (1993, p. 69) stated that "biota play *fundamental roles* in Earth's various systems, including soil-slope systems." I hope to show that biogeomorphology, and more specifically zoogeomorphology, is *both* "fun" and indeed fundamental to a thorough understanding of earth-surface processes and landforms. My approach is that of a geomorphologist trained as a physical geographer.

Definitions, scope, and limitations

Several of Thornbury's (1969) concepts of geomorphology illustrate the importance of knowledge of the past; yet at the same time, from a geomorphic perspective, most landscapes postdate the bulk of the earth's history. In this book, I examine only the late Holocene and its geomorphic processes associated with animals. It would be fascinating to speculate on the geomorphic

Figure 1.1. Mountain slope excavated by grizzly bear (*Ursus arctos horribilis*) in search of ground squirrel. Note 49-mm lens cap and pen for scale.

Figure 1.2. Trampled soil and grizzly bear pawprint, Otatso Creek drainage, Glacier National Park, Montana. Lens-cap diameter is 49 mm.

influence of Pleistocene megafauna, such as woolly mammoth wallowing (and, indeed, Viles [1988c] briefly spoke to this topic); or on the trampling effects of a herd of megadinosaurs such as *Brachiosaurus*. Without modern-day observations or a sufficient historical data base, however, such specula-tions must necessarily remain outside the confines of this book.

Viles (1988a, p. 1) defined *biogeomorphology* as "an approach to geo-morphology which explicitly considers the role of organisms." Subsequent-ly, I specifically defined *zoogeomorphology* as the study of the geomorphic effects of animals (Butler 1992). By animals, I mean both endothermic (warm-blooded) and ectothermic (cold-blooded) vertebrates as well as inver-tebrates. I exclude from this definition the anthropogenic changes brought about by human use of, and disruption of natural geomorphic processes on, the landscape. Excellent coverage and reviews of this topic already exist (Nir 1983; Trimble 1990; Phillips 1991; Sandgren and Fredskild 1991; Gou-die 1993). Along similar lines, this book does not examine in any detail the role of animals kept for agricultural purposes, but readers interested in that topic are referred to Trimble (1988). The emphasis here is on free-ranging, natural populations of animals (at least, as natural as possible in such a human-impacted world). Animals that have been introduced and become established in previously unoccupied ranges are, where appropriate, also dis-cussed.

Because they have been extensively studied in their own right from both geomorphic and biologic perspectives, I exclude coral reefs from inclusion in the topics under discussion in this book. Readers interested in these bio-genic landforms (in which only deposition occurs zoogeomorphically; ero-sion and transportation of coral sediments are carried out by other geomor-phic agents) are referred to Guilcher (1988), Spencer (1988), Viles's paper (1988b) in her edited volume (Viles 1988d, and references therein), and the aforementioned journal, *Coral Reefs*.

Obvious difficulties exist in identifying and documenting ocean- or lake-floor geomorphic impacts by aquatic animals. Where such impacts have been identified and have lasting geomorphic influence, they are described in this book. The emphasis here, however, is on terrestrial animals. Vertebrates are examined more closely than invertebrates, not because of any real or imagined hierarchy of importance to geomorphic processes, but rather be-cause of personal experiences and interests. In addition, the role of inverte-brates has received a larger degree of synthesizing attention in recent liter-ature, and readers interested in greater detail are referred to those works where appropriate.

The geomorphic effects of animals as defined in this book encompass the roles of animals in eroding, transporting, and/or depositing or causing the deposition of rock, soil, and unconsolidated sediments. At times it is difficult to distinguish *bioturbation* (mixing by biological means; see Hole 1981, for a thorough review of this topic) or *pedoturbation* (the mixing of soil by animals, also known as *faunalpedoturbation;* Johnson et al. 1987; Johnson 1989, 1990), from true geomorphic effects. In this book's usage, bioturbation or pedoturbation involves no net erosion from a site, whereas zoogeomorphic effects specifically involve the movement of rock, soil, or unconsolidated sediment from one location to another. Zoogeomorphic effects necessarily, then, involve erosion and deposition, whereas bioturbation or pedoturbation does not. Faunalpedoturbation may, however, give rise to several distinctive in situ landforms that must be examined.

Subsequent chapters in this book provide specific examinations of the geomorphic effects of invertebrates (Chapter 2), ectothermic vertebrates (Chapter 3), birds (Chapter 4), and mammals. Because of their widespread influence and variety of geomorphic influences, mammals are examined in more depth in several specific chapters. Chapter 5 addresses the geomorphic effects of digging for, and caching of, food (see Fig. 1.1). Chapter 6 examines the intertwined effects of wallowing, trampling, and geophagy (Fig. 1.3). The geomorphic effects of burrowing and den construction are examined in Chapter 7, which also includes a discussion on the nature and origin of Mima mounds, particularly as they relate to burrowing by rodents. Chapter 8 focuses on the geomorphic role of the beaver, an animal whose constructional efforts (Fig. 1.4) profoundly influence local hydrology and sediment budgets (Fig. 1.5). Chapter 9 concludes with a statement regarding the significance of animals as geomorphic agents, comments on human interference with those processes, and a hope for the future. This final chapter is followed in turn by an extensive reference list for readers interested in pursuing individual topics in more depth or from original sources.

Figure 1.3. The Walton Goat Lick (see Figure 6.8) occupies the gray area exposed in the river cutbank in center of photo; Middle Fork of the Flathead River is in lower foreground.

Figure 1.4. Typical small beaver dam constructed of sticks of wood, near East Glacier Park, Montana.

Figure 1.5. Stream flows through sediment that had accumulated in a beaver pond, now drained. The dam was approximately in the position of the camera taking the photograph. The depth of accumulated sediment is approximately 20–22 cm.

2
The geomorphic influences of invertebrates

The geomorphic effects of invertebrates are apparent to even the most casual observer who has seen an ant mound or earthworm castings. Invertebrates as geomorphic agents have been examined in a number of recent studies, but far fewer than those that examine their pedoturbational role. The papers by Goudie (1988) and Mitchell (1988) provide comprehensive review of the geomorphic role of select invertebrates, especially earthworms and termites. I do not attempt to duplicate the success of those papers here, as the emphasis of this book is on the geomorphic influences of vertebrates; rather, it is my intent in this chapter to summarize the geomorphic processes associated with invertebrates, and to present select examples of the geomorphic features they produce. In keeping with the emphasis outlined in Chapter 1, I concentrate on terrestrial invertebrates and offer only a brief survey of coastal and marine invertebrates.

Geomorphic effects of terrestrial invertebrates

Many insects, arachnids, and worms, as well as some crustaceans, are burrowing animals. Because burrows are underground phenomena, it is easy to believe that invertebrate burrows and the quantity of sediment they displace comprise small numbers, except in cases where material from underground has been deposited on the surface, as in the case of ant mounds (Fig. 2.1). The following sections illustrate the impressive nature of geomorphic work that a variety of invertebrates are capable of accomplishing.

Earthworms and termites

Because of the publication of excellent summaries on the pedoturbational effects of earthworms and termites, the emphasis here is on their direct and

indirect geomorphic effects. Readers interested in those pedoturbational aspects are urged to refer to the papers by Mitchell (1988) and Goudie (1988),
for temperate and tropical environments, respectively.

Earthworms and termites can contribute to soil erosion and denudation in
three major ways (after Mitchell 1988):

1. by removing the plant cover;
2. by digesting or removing organic material that would otherwise be incorporated into the soil and enhance soil stability; and
3. by bringing to the surface fine-grained material for subsequent wash
   and creep action.

Earthworm casts and termite mounds (see Fig. 2.2) are the surface manifestations of the latter action.

Earthworms

Charles Darwin produced the first detailed description of the pedoturbational actions of earthworms (Goudie 1988; Mitchell 1988), but little work on
the geomorphic effects of worms has followed. Rather, most research has
focused on quantifications of earthworm bioturbation rates from locations
around the world in comparison to the rates suggested by Darwin.

In effect, the geomorphic effects of earthworms can be broken down into
two categories: *direct* effects through cast production (microlandform creation), and *indirect* effects through influence if not outright control of processes of infiltration, soil creep, surface wash, and rainsplash detachment.
Goudie (1988) reviewed the phenomenon of earthworm-cast production,
and provided some quantitative data concerning the amount of soil involved. His review suggested a typical range of 2–30 kg of soil m^{-2} yr^{-1},
significantly higher than the 1.5 kg m^{-2} yr^{-1} reported from Luxembourg
soils by Hazelhoff et al. (1981). More limited data from Goudie's review
provided estimates of the overall surface denudation rate, with a range of
3.1–25 mm yr^{-1}. Most studies he examined suggested that the higher end of
the latter range is probably more typical.

Earthworms' burrowing and their contribution to a soil's overall biomass
strongly influence the indirect effects of processes of infiltration, creep, rainsplash detachment, and surface wash (A. Meadows 1991; P. S. Meadows
1991; Meadows and Meadows 1991b), as do their consumption of vegetation at a site. In many soils, earthworms represent the dominant fraction of
soil biomass (James 1991), the influence of which is seen in the ability of
decomposed biomass to reduce erodibility through humus generation (Jun-

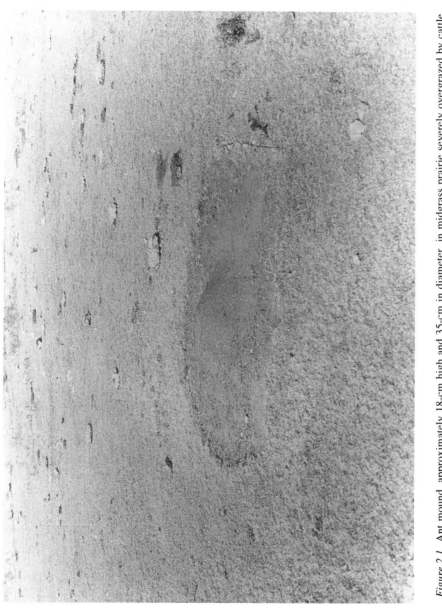

Figure 2.1. Ant mound, approximately 18-cm high and 35-cm in diameter, in midgrass prairie severely overgrazed by cattle, western Kansas. Note the barren perimeter zone around the ant mound, and the numerous cattle-manure deposits.

gerius, Van den Ancker, and Van Zon 1989). The burrowing in, and trans-
formation of, the soil by earthworms also changes a soil's infiltration capacity (Goudie 1988; Mitchell 1988; Edwards et al. 1990; Robinson, Pearce,
and Ineson 1991; Schrader and Joschko 1991). Goudie (1988) noted that
soils with earthworm activity over an 8–10-yr period had infiltration capacities of 4.6–6.4 m per day, whereas soils without earthworms had infiltration
capacities of <0.05 m per day. The obvious indirect result of the latter would
be increased surface wash and erosion, whereas the presence of a high infiltration capacity allows a pronounced reduction in surface runoff, in turn resulting in lower fluvial erosion. Edwards et al. (1990) corroborated such an
interpretation with their study of cornfields in Ohio. When crop residues
were left on the surface using no-till corn-management processes, earthworms thrived. Soil physical properties were dramatically altered by the
presence of numerous earthworm burrows, and infiltration into the surface
via those burrows accounted for an average of 4% of rainfall, thirteen times
more infiltration than was predicted based solely on a soil's cross-sectional
area.

In addition to the effects of earthworms produced by burrowing and cast
production on the surface, there has been recent contradictory research presented on the effects of vegetation removal from the surface by earthworms.
Hazelhoff et al. (1981) showed that *Lumbricus terrestris* earthworms in a
forested Luxembourg watershed effectively strip a surface of its leaf cover
by pulling leaves down into underground worm burrows. The removal of
leaf litter exposes the bare forest floor to splash erosion or erosion by overland flow. They noted that this activity varied seasonally, but concluded
(p. 249) that the importance of earthworms "in erosion lies in the fact that
they create bare soil, and thus promote erosion," and in comparison to moles
in the same area "created a greater potential for erosion under deciduous
forest in the area and during the period of study" than the moles.

In a subsequent study in Luxembourg, Jungerius et al. (1989) also examined the role of earthworms in leaf-litter removal and its influence related to
subsequent erosion. They noted that, when comparing fine-grained soils to
sandy soils, leaf consumption by earthworms did exacerbate erosion because
more worms were present in the fine-grained soil and therefore exposed a
greater surface area than in the sandy soil. However, they further noted that
worms reduced a soil's erodibility by improving its structure (and hence infiltration capacity) and by increasing the organic matter content. They also
suggested that on surfaces stripped of leaves by worms, the resultant surface
earthworm casts help soil particles adhere and resist erosion. Basically,
then, the greater an area stripped of leaves by worms, the more one might

expect an increase in surface wash and runoff, *except* that such soils, because of their inherent richness and improved structure and adhesion, are inherently more resistant to erosion! Clearly, additional research is necessary
to resolve the question of the indirect erosional effects of leaf-litter stripping by earthworms.

Termites

As Goudie (1988) pointed out, termite mounds or hills (Fig. 2.2) are the
most impressive manifestation of termite activity; indeed, a termite mound
graced the cover of Viles's (1988d) book. Termite mounds appear in the
fossil record at least as far back as the Miocene (Bown and Laza 1990).
Termites are distributed in most climatic zones warmer than those of tundra
regions, but they reach their greatest visibility in the tropical environment.
Goudie (1988) provides a table illustrating the range of heights of termite
mounds reported in the literature. In general, those values range from 0.05
to 7 m. As Branner (1909, p. 482) describes, termite mounds "are rudely
domed, rounded or conical, and the method of adding to the outside gives
them a bumpy, lumpy appearance, so that . . . they resemble gigantic Irish
potatoes." Individual termite mounds may contain >15 m^3 of earth, but such
cases are exceptional (Branner 1909). Mound densities range from fewer
than 1 to over 850 per hectare (Pomeroy 1977, 1978; Pullan 1979; Goudie
1988), but San Jose et al. (1989, p. 353) pointed out that the "pattern of
termite mounds in each of the different physionomic [*sic*] types of savanna is far from uniform; the savannas may be densely populated or have
no mounds at all." In local areas, the distribution of mounds ranges from
random to the "regular" side of random (Fisher 1993). In some cases, the
widespread distribution and size of the mounds produce Mima-like mounds
similar to those associated with burrowing mammals (Lovegrove 1991) (see
Chapter 7). Necessary site characteristics amenable to mound formation include soils with a relatively high water-holding capacity or low water loss, a
deep soil profile with lower layers possessing high water content, high clay
and low sand content, and little disturbance from human or large-animal
activity (Arshad 1981; Lacher et al. 1986; Goudie 1988; Thomas 1988; San
Jose et al. 1989; Whitford, Ludwig, and Noble 1992).

The longevity of termite mounds varies considerably, with some nest-
mound sites occupied on the scale of centuries (Goudie 1988). Erosion of
mounds is enhanced by anteaters and antbears searching for termites for
food (see Chapter 5), by geophagy of salt-rich mounds (see Chapter 6), and
by compression as animals sit on or rub against the mounds (Goudie 1988).

Pullan (1979) and Goudie (1988) described and summarized the literature on miniature wash pediments that surround termite mounds and owe their sediment to erosion of the mounds. In extreme cases, such wash pediments may extend up to 60 m in diameter away from the mounds (Goudie 1988).

The other major effect caused by termites is the construction of "covered runways or 'sheetings' on the ground surface and on vegetation" (Goudie 1988, p. 181). These are essentially analogous to earthworm castings, and Goudie (1988) provided the following summary data of the quantities of material translocated out of the ground and onto the surface for sheeting formation: in Nigeria, 300 kg ha^{-1} yr^{-1}; in Senegal, 675–900 kg ha^{-1} yr^{-1}; and in Kenya, 1,059 kg ha^{-1} yr^{-1}. Goudie noted that these rates are "significant," but on an order of magnitude less than that of cast production by earthworms. Even combining the amount of soil involved in the formation of both mounds and sheetings results in values lower than those associated with earthworm casts.

Ants

There are approximately twelve thousand living species of ants (Sudd and Franks 1987), but many of these are aboveground or tree-dwelling species that have no geomorphic or pedoturbational significance. This section deals only with the role of ants as geomorphic and/or pedoturbational agents, and their associated surface landforms, that is, ant mounds.

Effects on soil

Several papers have summarized the general effects of ant burrowing and nesting (Petal 1978; Elmes 1991; Woodell and King 1991), and readers are referred to those papers for specific information. Elmes (1991) noted that many burrowing ants do not actually raise mounds above the surface, but instead occupy widespread underground tunnels and "roadways." The geomorphic significance of these ants rests in their aeration of the soil and the subsequent creation of soils that allow rapid drainage and throughflow.

Physical and chemical changes associated with soils subjected to ant burrowing and tunneling specifically include lowering of soil bulk density and increasing soil porosity (Mandel and Sorenson 1982), increasing levels of organic matter, alteration of soil pH (in alkaline soils the pH of ant sites is decreased, whereas in acid soils it increases; Petal 1978), and increasing nutrient levels (Elmes 1991). A number of specific case studies also address these issues. Laundré (1990) described the moisture-retentive capabilities of

Figure 2.2. Photograph of a termite mound, similar to those in Thomas (1988).

soils beneath ant mounds in semiarid rangeland in Idaho, and Friese and Allen (1993) illustrated organic matter and mycorrhizal fungi enrichment in similar soils of mounds of the western harvester ant (*Pogonomyrmex occidentalis*) in the semiarid shrub steppes of the Great Basin region of the United States. Soils of South African ant mounds studied by Dean and Yeaton (1993) had higher infiltrability and significantly more organic matter, phosphorus, potassium, and nitrogen than soils of intermound areas. Similar results were described by Mandel and Sorenson (1982), Carlson and Whitford (1991), and Blom et al. (1994) for mound soils of the western harvester ant in Colorado, New Mexico, and Idaho, respectively. Conversely, however, Culver and Beattie (1983) found lower levels of micronutrients on ant (*Formica canadensis*) mounds versus nonmound soils in a Colorado montane meadow, but these results could reflect the moister and cooler microclimate of their study site.

The pedoturbational effects of ants has been described in detail by a number of researchers. Johnson and coworkers recently examined the geomorphic influences of ants in a series of papers examining the genesis of Quaternary stone lines in soils (Johnson et al. 1987; Johnson 1990; Johnson and Balek 1991). They attributed some "biomantles" (differentiated zones in the upper part of soils produced largely by bioturbation) directly to the faunalturbational effects of ants. Johnson et al. (1987) described one case where earthworms and ants working in concert buried an ungrouted brick patio several centimeters in under twenty-three years, and stepping stones were buried to a depth of 4.8 cm in less than forty-eight years. Other papers examining the pedoturbational influences of ants include those of Baxter and Hole (1967), Mandel and Sorenson (1982), and Dean and Yeaton (1993). Baxter and Hole, who examined soils pedoturbated by ants in southwestern Wisconsin, concluded that the work of ants is important in the maintenance of a high content of clay in the upper A horizon of the prairie soil, and in the thickening of the root-filled A1 horizon at the expense of the B horizon (Baxter and Hole 1967, p. 428).

Ant mounds

Ant mounds are the most visible manifestations of the pedoturbational activities of ants, and have been the subject of numerous studies that describe their morphology and density. Because of the diversity of species of ants involved in production of mounds, it is difficult to generalize about mound size and density. I therefore instead describe a variety of studies as representative of the range of ant-mound morphometrics, and search for commonalities.

The work of Branner (1909) was one of the first scientific examinations of ant-mound morphometry and density. His paper examined ants of tropical Central and South America, particularly in Brazil. Branner clearly differentiated ants and their mounds from termites, and described mounds as high as 5 m, with bases 15–16 m in diameter, each therefore containing ~340 m^3 of earth. Sketches from photographs in his paper clearly illustrate the massive nature of these mounds. Branner also described underground passages, varying from 1 to 5 cm in diameter, that were over 300 m long.

Ant mounds in the central United States are substantially smaller, yet still displace impressive quantities of material from below. Baxter and Hole (1967) described a density of over 1,500 mounds per hectare in southwestern Wisconsin, each of which had a volume of 0.02 m^3 (37 cm diameter × 15.3 cm height). The mounds were estimated to occupy 34 m^3 ha^{-1}, comprising approximately 1.7% of the total ground surface.

Western harvester ant mounds in the semiarid western United States (see Fig. 2.1) have been the subject of several pedogeomorphic studies, including those of Scott (1951), Mandel and Sorenson (1982), Laundré (1990), Carlson and Whitford (1991), and Clark (1993), the last of whom described an atypical double-mound scenario. Scott (1951) noted that ant mounds were visible on aerial photographs taken over Wyoming, occupying up to 2% of the total surface of an area. (The use of aerial photographs for inventorying mound populations is addressed in Baroni-Urbani, Josens, and Peakin [1978].) He estimated that over six million mounds occurred in the Wind River Basin alone, and his descriptions of them remain a standard to which other workers refer. He describes mounds such as those seen in Figure 2.1 as being in the shape of a cone, with a slope angle of about 27°. Variations in the angle of repose he attributes to the particle size of materials used in mound construction. The mounds are typically situated near the middle of a circular clearing, created by the ants, of about 3–4 m in diameter. The clearings around the mound expose the surface soil to secondary erosion by wind.

Mandel and Sorenson (1982) described similar harvester ant mounds from a semiarid valley in central Colorado. They noted the mounds' typical cone shape and the conspicuous circular clearings around them. Measuring twenty such features, they found an average clearing diameter of 2.2 m, and an average mound volume of 32 l, with a range of 14–51 l. The mounds themselves occupied only about 0.10% of the ground surface, but inclusion of the denuded areas circling the mounds gave a figure of 2% of the land area affected. Mean particle-size distribution of the soil material comprising the mounds was 77% sand, 15% silt, and 8% clay, compared with 49% sand, 32% silt, and 19% clay in the upper 20 cm of undisturbed soil.

Results similar to those of Scott (1951) and Mandel and Sorenson (1982) were described by Laundré (1990) from semiarid rangeland in Idaho, and by Carlson and Whitford (1991) from the forest-steppe ecotone near Los Alamos, New Mexico. Laundré (1990), although primarily interested in moisture patterns under harvester ant mounds, reported a density in excess of sixteen mounds per hectare. Carlson and Whitford (1991) reported that the denuded zones around western harvester ant mounds occupied about 1–1.2% of the total surface area. Mounds occurred in densities of fourteen to seventeen per hectare and had an average mass of 38–48 kg. Coarse particle sizes similar to those described by Mandel and Sorenson characterized the New Mexico mounds.

Brian (1978) and Elmes (1991) summarized several studies of ant-mound morphometry and density. Elmes noted a particular case from a tidal meadow of a Baltic island, where ant (*Lasius flavus*) mounds occupy 7–12% of the land area at a density of 2,500 mounds per hectare. Each mound contained ~40 *l* of soil, for a total estimate of 100 m^3 of mound soil per hectare. Elmes also noted that similar figures could be calculated for *L. flavus* sites in Britain, and Petal (1978) stated that the mounds of this ant occupy 10–11% of areas with substrate comprised of silty gravels. Woodell and King (1991) also described the effects of *L. flavus* on British lowland sites.

In a study conducted in the Ardeche district of southern France, Aalders, Augustinus, and Nobbe (1989) examined ant mounds in cultivated areas, rather than in natural areas as was typical of the studies cited above. They report a total of 133,541 ant mounds in their 782-ha study area, or about 171 mounds per hectare. Mounds with diameters >10 cm had average volumes of 170 g and comprised 78% of the total mound population. They conclude that ant activity on soil erosion did not appear to be significant, but note that their results apply only to intensely cultivated areas.

Mitchell (1988), summarizing research on ant mounds in Australia, concluded that although their mounds were in some cases quite conspicuous, their overall importance as agents of pedoturbation was greatly exaggerated. He noted (p. 48) that this situation did "provide a salutary lesson on the risk of exaggerating the importance of conspicuous phenomena," that is, ant mounds.

In contrast to the viewpoint of Mitchell (1988) expressed above, Cox, Mills, and Ellis (1992) described Mima-type mounds (also see Chapter 7) of up to 1.5 m high and ~20 m in diameter from a seasonally waterlogged grassland area in Argentina where fossorial rodents are absent, and attributed their development to the accumulative effects of fire-ant mound construction. The ants construct nest mounds about 20 cm in height and 60 cm in diameter; subsurface tunneling causes horizontal soil translocation that gradually results in soil accumulation and the growth of large earth mounds.

In the subarctic and arctic environments, ants and ant mounds are noticeably absent in large number. However, in unusual circumstances, local ant populations and geomorphic influences may be high. Luken and Billings (1986) described the degradation of hummocks formed by cryoturbation in a subarctic Alaskan peatbog, and attributed the hummock instability, degradation, and collapse to the tunneling of ants. Apparently, the warm, dry microclimate of senescent peat hummocks is conducive to ant colonization, which in turn leads ultimately to hummock collapse, microhabitat cycling back to permafrost conditions, and a new generation of peat hummocks.

Ant denudation rates

Denudation rates associated with ants have been described since soon after Charles Darwin's classic description of soil turnover associated with earthworm activity. Branner (1909) compared Darwin's statement regarding the quantity of earth brought to the surface by earthworms in many parts of England (10,516 kg of soil per acre annually) with results from his study of ants in Brazil. There, he found the total weight of earth surfaced by ants over 1 ha in a hundred years to be 3,226,250 kg, compared to the (metrically converted) 2,598,500 kg ha^{-1} described by Darwin. Branner noted, however, that the levels of earth brought to the surface in Brazil were "exceptional."

Baxter and Hole (1967) summarized studies of ant pedoturbation in the upper U.S. Midwest. They concluded that ants move about 6 ha-cm of soil material to the surface in one to three centuries.

Woodell and King (1991) described a study site in south Wales where one area of a thousand square meters contained more than nine hundred mounds of *Lasius flavus*. They estimate that the largest mounds increase in volume by about 1 l yr^{-1}, so that soil is deposited over the entire study area at a rate equivalent to 0.5 mm yr^{-1}. They compared this result to that of Darwin, where from observations in several different grasslands it was calculated that earthworms deposit 2.1–5.6 mm of soil per year, a rate only four to eleven times faster than ants. When combined with the estimates of surface coverage of ant mounds and associated vegetational clearings of ~2–12% of a given area, it would seem that Mitchell's (1988) comment about the exaggerated importance of ants as geomorphic and pedoturbational agents is itself flawed. Ants produce widespread visual evidence of their geomorphic abilities, are spatially widespread throughout the world, and move vast quantities of soils and substrate materials in their tunneling and mound-building activities.

Other terrestrial invertebrates

The geomorphic effects of other terrestrial invertebrates are relatively minor in comparison to the work of ants and termites. Nevertheless, several interesting effects bear comment in the sections to follow.

Insects, arachnids, and gastropods

Other invertebrates that accomplish minor amounts of geomorphic work include insects, arachnids, crustaceans, gastropods, and burrowing meiofauna (protozoans and metazoans; Reichelt 1991). Geomorphic accomplishments by insects include those of the mole cricket (*Gryllotalpa hexadactyla*), which is recorded in the fossil record back to the Permian (Metz 1990), cicadas, scarabs, and other beetles. Mitchell (1988) described his research with Humphreys in Australia, where they illustrated minor bioturbational effects of cicadas at rates ranging from 0.03 to 0.2 t ha^{-1} yr^{-1}. Mole crickets construct permanent burrows in moist, cohesive clays into which eggs are laid. Mole cricket burrows as deep as 30–75 cm have been reported in the literature (Metz 1990). Similar methods and tunnel forms have also been reported for certain beetles (Clark and Ratcliffe 1989) and scarabs (Evans 1991). Kalisz and Stone (1984) reported that scarabs possess the ability to bioturbate soil at a rate of 2.61 t ha^{-1} yr^{-1}.

Some bees (Family Anthophoridae) also dig burrows, typically into sandy or silty stream-bank sediments (Seely, Zegers, and Asquith 1989). Digger bee burrows may penetrate into vertical banks of sediment to distances of at least 15 cm. The tree lizard (*Urosaurus ornatus*) may subsequently occupy and expand these burrows, apparently for thermal protection and safety from predation (Seely et al. 1989).

Many spiders are known to live in small burrows or holes in the ground, but very little information exists regarding the geomorphic efforts of arachnids. Polis, Myers, and Quinlan (1986) described the burrowing biology of desert scorpions, but offered little insight into the geomorphic effects of such burrowing. Formanowicz and Ducey (1991) reported on the burrowing behavior and soil manipulation by tarantulas (Fig. 2.3) (*Rhechostica hentzi;* Araneida: Theraphosidae) collected from grassland habitats in Texas. Al-

Figure 2.3. The tarantula (*Rhechostica hentzi*), a burrowing arachnid, is common in the southwestern United States and northern Mexico; near Chaco Canyon, northwestern New Mexico.

though they provided no information on the amount of sediment displacement associated with tarantula burrowing, Formanowicz and Ducey reported that large, naturally occurring clumps of soil were carried by claw, or dragged in a silk–soil mass, as far as 52 cm away from burrow entrances. Even accepting the limited amount of sediment involved, it is nevertheless clear that arachnids must be considered as agents of pedoturbation and sediment transport.

Little is known about the role of terrestrial gastropods as geomorphic agents. Viles (1988b) examined organisms associated with bioerosion in karst landscapes, and commented briefly on the role of snails. She noted that snails may produce tubular boreholes into limestone up to 50 mm deep, and recognized a developmental sequence leading from shallow scrapes through simple holes, to complex holes and rock honeycombs. The presence of snail holes in limestone quarries only two hundred years old suggested a rate of formation of up to 0.15 mm yr^{-1}.

Crayfish

Crayfish are terrestrial crustaceans found in low-lying terrain throughout the Gulf Coast region of the southern United States, as well as in such diverse habitats as Indiana and Australia (Mitchell 1988). Many crayfish species dig deep vertical and complex shafts leading to long-lasting underground burrows. Crayfish burrowing has occurred as a geomorphic and pedoturbational activity since at least the Triassic period (Hasiotis and Mitchell 1993; Hasiotis et al. 1993). Burrows are complex and may be as long as 4–5 m (Hobbs 1981; Hasiotis, Mitchell, and Dubiel 1993). Hobbs (1981) identified three categories of crayfish burrowers and burrows based on the amount of time spent in the burrow, its architecture, and its connection to open water:

Primary: Crayfish construct burrows with complex architecture, unattached to open water, and spend the majority of their life in the burrow. Burrows are vertical, obtain great length, and branch out horizontally below the local water table (Hobbs and Whiteman 1991; Hasiotis and Mitchell 1993; Stone 1993);

Secondary: Crayfish construct burrows that are attached to open water and exhibit some branching and chamber development, and spends much time out of the burrow (Hobbs 1981; Hasiotis and Mitchell 1993); and

Tertiary: Crayfish spend most of their lives in open water, burrowing to reproduce or escape desiccation (Hobbs 1981).

Here, the work of primary burrowers is of importance because of its great geomorphic and pedoturbational ramifications.

The digging of burrows by crayfish results in the deposition of surface mounds, also known as crayfish "chimneys" or "turrets," across the landscape (Fig. 2.4). The mounds are surface entry points to a tunnel system frequently over 1 m in depth and chiefly 4–8 cm in diameter (Stone 1993). In some cases the mounds, frequently composed of clay, become hardened by the sun so that they are an impediment to farm machinery. Burrows may so permeate the subsurface that farm machinery and livestock cause ground collapse into the underlying burrow system (Hobbs and Whiteman 1991; Stone 1993). The pedoturbational activities in burrow excavation and mound emplacement also bring an excess amount of sodium salts to the surface in some locations, impeding further agricultural usage of the area (Hobbs and Whiteman 1991).

The morphology of crayfish mounds has been described in some detail by Hobbs and Whiteman (1991). They reported densities of 2,730 mounds per hectare from one site in eastern Texas, with average mound heights of 12 cm and diameters of 28 cm. They went on to report that >17,000 kg of soil and 40 kg of sodium per hectare were brought to the surface each year, such that the entire surface "must have exhibited some stage of mound construction or erosion over a period of <3 years" (p. 128). Yet these values were not the highest cited in their report! At six other east Texas study sites the estimated number of mounds per hectare was 44,124 (averaging 7 cm tall and 15 cm in diameter), 63,505, 31,214, 62,676, 61,775 (covering 15% of the surface area), and 42,470 (with >40,000 kg of soil moved per hectare). Individual mounds examined weighed up to 40 kg, although 11–25 kg seemed to be the normal range.

The density of crayfish burrows and tunnels in areas of the southern Gulf Coast of east Texas, western Mississippi, and eastern Alabama is so great that crayfish play a major role in the hydrology of the region. Stone (1993, p. 1,099) asserted that, in poorly drained areas, "[w]hen water tables are near the surface, however, the network of large-diameter crayfish tunnels . . . allows lateral flow at rates unlikely to be equaled by the activity of any other burrowing organism," and noted that the effectiveness of crayfish in "mixing the surface of a very poorly drained soil seems comparable to that of large earthworms such as *Lumbricus terrestris* L. (alone or together with burrowing predators)." The bulk of the research on crayfish, however, has been carried out by soil scientists and zoologists, rather than geomorphologists. Further research efforts by geomorphologists are awaited to integrate fully the geomorphic effects of crayfish into a hierarchy of significance in which termites, ants, and earthworms are at or near the pinnacle.

Figure 2.4. A crayfish chimney on the floodplain of the Roanoke River, eastern North Carolina. Lens-cap diameter is 49 mm.

Aquatic invertebrates

Recent research has examined and defined the importance of the bioturbational role of aquatic invertebrates in intertidal (Cadée 1976, 1979; Miller 1984; Spencer 1988; Viles 1988b; Akpan 1990; Andersen and Kristensen 1991; Meadows and Meadows 1991a,b; Dworschak and Ott 1993) and benthic (De Wilde 1991; Meadows and Meadows 1991b) habitats; however, it is not clear from most studies if the sediment mixing created by animals such as sponges, gastropods, crustaceans, and other organisms has any lasting *geomorphic* effects. This section briefly examines actual geomorphic results produced by aquatic invertebrates; readers interested in bioturbational activities and rates are referred to the works cited above.

By far the most comprehensive review of the geomorphologic effects of subtidal and intertidal invertebrates and associated erosion rates is that of Spencer (1988), who quantifies the contribution to erosion by a diversity of coastal invertebrates. Other works include that of Fischer (1990, p. 314), who examined the Pacific coast of Costa Rica and described the importance of invertebrates in producing "typical morphological structures of the coast . . . of two different orders of magnitude." He identified large-scale morphologies associated with notch formation at mean low-water level, and small-scale morphologies associated with borings (of sponges [only in carbonates], polychaetes, sipunculids, bivalves, sea urchins, and pistol shrimps) and grazing tracks (produced by gastropods and polyplacophores) in tidal pools. Similar small-scale morphologic features have been reported from other coastlines by Spencer (1988), and specifically for sponges by Rützler (1975), and Bromley and D'Alessandro (1990). The latter noted that sponge borings dominate all samples of recent and fossil coral material from deepsea environments in the Mediterranean Sea. Rützler (1975) quantified the contribution of sponge boring to coral mud, noting that sponge-generated erosional chips can comprise >40% of such mud, and estimated the burrowing potential of sponges in limestone substrate as 256 g m^{-2} yr^{-1}. Spencer (1988) reviewed the literature on bioerosion by sponges and polychaetes on coral reef communities, and reported carbonate-removal rates ranging from values similar to those of Rützler up to about 8 kg m^{-2} yr^{-1}, with an extreme value of 20–25 kg m^{-2} yr^{-1} for a transplanted block of material in Bermuda. Sponge erosion rates typically ranged higher than those of polychaetes, but values in the range of 4–4.75 kg m^{-2} yr^{-1} were reported for polychaete erosion at Lizard Island, part of Australia's Great Barrier Reef.

As noted, Fischer (1990) examined the position of the mean low-water level and an associated erosional notch found along the entire Pacific coastline of Costa Rica. He attributed the creation of this notch, described as a few tens of centimeters in width, to "biodestruction" associated with boring invertebrates, particularly sea urchins. Spencer (1988, p. 261) also noted the creation of bioerosion notches attributable to " 'biological corrosion,' a solution process undertaken by microorganisms and some macroborers which modifies the substrate but provides no erosion product, and 'biological abrasion,' a series of physical processes carried out by grazing, burrowing and boring organisms which result in particulate debris production."

Spencer (1988) also provided a detailed analysis of the role of benthic invertebrates in sediment destabilization and microlandform creation on the seafloor. Rather than simply reiterate those comments here, I focus on his description of the sediment-processing activities of callianassid shrimps. These shrimps are burrowing animals that create funnel-shaped depressions into the benthic sediments with corresponding "volcano-like mounds, up to 30 cm high and reaching densities of 10 mounds m^{-2}" (Spencer 1988, pp. 298–9). The depressions beneath these microlandforms may extend to depths of 2 m, although 50-cm depth is more typical. Photographs and diagrams of the shrimp mounds suggest a strong similarity to their nonmarine counterparts, the mounds and subsurface chambers produced by crayfish.

Conclusions

A good rule for summarizing the geomorphic influence of invertebrates would be "Just because you can't easily see them doesn't mean that they're not doing a great deal of work"! Whether on land, in intertidal pools and rocky coastal platforms, or in benthic sediments, invertebrates create a diversity of small-scale landforms. They bioturbate massive quantities of sediment, and their actions play major roles in determining the erodibility of a landscape. A great deal of room remains, however, for more accurate quantification of the amounts of actual sediment eroded and transported, versus simply bioturbated. Work could also focus on the durability of landforms such as ant mounds, crayfish chimneys, termitaria, and shrimp "pseudo-volcanoes," as well as on simply identifying other surface features created by invertebrates. Finally, the indirect effects of invertebrates – expressed through contribution to soils of biomass that inhibits their subsequent erosion, through vegetation stripping and its effect on rainsplash and surface wash, and through pedogenetic effects such as changes in infiltration capacity – could all provide areas for additional research for zoogeomorphologists for decades to come. As in other realms of science, Charles Darwin has pointed the way, and zoogeomorphologists must come to grips with how to carry that legacy to fruition.

3
The geomorphic accomplishments of ectothermic vertebrates

This chapter examines the geomorphic role of fish, amphibians, and reptiles. As stated in Chapter 1, I must regrettably ignore the geomorphic effects of dinosaurs, Permian-era mega-amphibians, and other geologically distant effects of the ectothermic (cold-blooded) vertebrates. Each of the groups of modern ectothermic animals does induce geomorphic change; however, the bulk of these changes are either transitory or aquatic in nature, and are therefore only briefly described.

The geomorphic role of fish

The geomorphic work of fish can be categorized into three major efforts: nest building/digging, disturbances associated with feeding, and burrowing. Nest creation is a relatively simple affair, in which a fish prepares a nest by cutting into the gravel of a coarse streambed with downward beats of the tail (Hansell 1984). The salmonids – that is, salmon and trout – are probably the best known of such nest builders. Von Frisch (1983), Chapman (1988), Crisp and Carling (1989), and Kondolf et al. (1993) describe the typical salmon nest, or "redd," as a hollow pit 10–20 cm deep and 1–2 m long, oriented in the direction of the stream current. The base of the pit, also called the "pot" (Crisp and Carling 1989), is comprised of coarser lag gravels (typical particle size for a brown trout redd in California was 45–64 mm, whereas cutthroat trout in Idaho spawn in gravel with particles up to 100 mm in diameter; Kondolf et al. 1993; Thurow and King 1994), and a "tail-spill" of finer-grained material forms a low mound immediately downstream of the pit (Lisle 1989; Kondolf et al. 1993). The size of the pot is positively related to the size, especially the length, of the female fish (Ottaway et al. 1981; Van den Berghe and Gross 1984; Crisp and Carling 1989). Cutthroat trout redds in Idaho averaged 1.58 m long by 0.60 m wide, with the pit covering 46% of the redd area (Thurow and King 1994).

After egg deposition and fertilization, the female salmonid covers the eggs under 10–40 cm of bed material with additional sweeps of the tail (Lisle 1989). The gravels and interstitial fines excavated during the redd construction are exposed to currents and differentially transported downstream: Gravels move only a short distance, but fine-grained sediments are swept further along (Kondolf et al. 1993). The net result of redd construction and the removal of fines by the current is a stream bottom with a different, coarser particle-size distribution than adjacent gravels not utilized for spawning (Chapman 1988). Kondolf et al. (1989) described paired samples of rainbow trout redds and mainstream gravels from tributaries streams of the Colorado River above Lee's Ferry. They found that spawning fish reduced the percentage of fine (<0.85 mm) sediment from 6% to 2% on average, increased the average particle size, and improved sorting. Spawning and redd construction therefore has both temporary and longer-term geomorphic effects: In the short term, a transient microtopography of pits and tailspills is created along the stream bottom; over a longer period, spawning sites on the streambed are maintained, while areas of gravel that are coarser than surrounding unused gravels are concentrated, and fine-grained sediments are transported downstream.

Unlike the case of the salmonids, nest building in other fishes is most typically carried out by the male (Hansell 1984). The three-spined stickleback (*Gasterosteus aculeatus*) is an example of a species in which the male excavates a shallow depression as the basis for a more complex nest comprised of collected vegetation. The sand goby (*Gobius minutus*) digs out a nest beneath shells, and in concert with this excavation also produces a small ditch or trench leading to the shell-protected nest site (Von Frisch 1983).

By far the most controversial, and unlikely, geomorphic effects attributed to the excavation of spawning beds by fish are the Carolina Bays of the southeastern U.S. Coastal Plain. These "bays" are shallow, elliptical, wetland depressions, typically characterized by a long-axis orientation of northwest–southeast and a prominent sand rim (Schalles et al. 1989; Lide 1991). Although the size of individual bays varies, many are at least 1 km long on their long axis. Over five hundred thousand such bays exist from southern Georgia to Maryland along the Coastal Plain (Price 1968). Although their origin is still open to debate, current theory suggests that they were created by southwest–northeast prevailing winds during the Pleistocene (Carver and Brook 1989). However, Grant (1945) attributed their creation to vast quantities of schooling, spawning fish creating tidal-flat excavations during the Pleistocene. Because of regional differences in the shape of the bays between the northern and southern populations, it was suggested that fish spawning in the north did so in elliptical shoals, "whereas their southern congeners favored oval-shaped schools" (Grant 1945, p. 444). No explanation for this geographic differentiation in the shape of the schools of fish was offered, but marine geologists suggest that whatever the shape, the patterns of subaerial barriers (sand rims) within which all bays lie would not have survived marine invasion at the end of the Pleistocene.

Bottom-feeding fish stir up great quantities of sediment in their search for food, but the geomorphic effects are largely ephemeral. An exception is the family of rays along the southeastern U.S. coast. The following discussion is a synopsis of the research of Howard, Mayou, and Heard (1977), carried out along the Georgia coast of the United States. There, three rays – the southern stingray (*Dasyatis americana*), the bluntnose stingray (*D. sayi*), and the Atlantic stingray (*D. sabina*) – produce widespread hole excavations in tidal flats and in estuarine sand bars and channels. Digging is accomplished by excavation with the pectoral fins or "wings" (although see Gregory et al. [1979] for a discussion of how some New Zealand rays excavate feeding depressions by jetting water through the mouth and/or gill clefts), and is primarily associated with periods of ebb tides. Ray feeding depressions vary in diameter from 6 cm to as much as 1 m in diameter, are typically ~6 cm deep, and are partially surrounded by a low sand rampart 3–4 cm high. Morphologically similar depressions associated with benthic feeding off the coast of California were described by Cook (1971), and attributed to the work of the California halibut (*Paralichthys californicus*), the bat ray (*Myliobates californicus*), and the common skate (*Raja erinacea*).

Although coral reef formation is beyond the scope of this book, it is worth mentioning that fish are major agents of bioerosion on reefs (Bardach 1961; Bromley 1978; Bromley and D'Allesandro 1990). Much of the research associated with feeding on, eroding, and breaking off of coral has been thoroughly summarized by Spencer (1988). Frydl and Stearn (1978) provided quantitative estimates of the amount of erosion associated with parrotfish (*Sparisoma viride*) in reef environments on the west coast of Barbados. There, rates of bioerosion associated with parrotfish feeding on coral were 61 g m^{-2} yr^{-1}, 40 g m^{-2} yr^{-1} on the bank reef, and up to 168 g m^{-2} yr^{-1} in one specific study area. Bardach (1961) also stressed the importance of parrotfish in the degradation of coral reefs. In his study of a typical Bermudan coral reef, he calculated that omnivorous reef fish ingested and redeposited sand, coral scrapings, and other calcareous materials in the amount of at least 2,300 kg ha^{-1}. Parrotfish and surgeonfish (Acanthuridae) were the most important fishes, with consistently large amounts of sand, coral fragments, and limestone powder present in their digestive tracts.

Burrowing by fish is a significantly greater geomorphic process than either bottom feeding or nest digging/building, and is a process that has been around since at least the late Paleozoic era (Carlson 1968). In some cases, burrowing is simply a matter of the fish covering itself with inorganic sediment and shell fragments as a means of protection and camouflage; in such cases, "burial" is probably a more precise term for the action than "burrowing" (Atkinson and Taylor 1991). In other cases deep burrows and trenches are excavated on lake or sea bottom, in freshwater, brackish, and saltwater environments.

The spatial concentration of burrows can be quite high: The mudskipper (*Boleophthalmus dussumierei*) lives on coastal mud flats and excavates 1–2-m-deep vertical burrows in such close proximity (each fish occupies a territory of ~1.65 m^2) that pentagonal polygonal ground results, with the boundaries of each vertical shaft clearly marked by mud ramparts constructed from mouthfuls of mud scooped from burrows (Hansell 1984). Jones et al. (1989) reported a density of 0.3 yellowedge grouper (*Epinephelus flavolimbatus*) burrows per 1,000 m^2, and many of these were "trench" burrows 7–8 m long, 2–3 m wide, and 1–1.5 m deep. Using these ranges, a large yellowedge grouper burrow displaces 14–36 m^3 of sediment on the ocean floor. Able et al. (1982) described the burrows produced by tilefish (*Lopholatilus chamaeleonticeps*) in the Hudson Submarine Canyon off the east coast of the United States, where tilefish abundance was estimated at about 660 km^{-2}. Larger burrows were up to 4–5 m in diameter and at least 2–3 m deep, so that even if only a small percentage of the 660 fish are large adults, considerable sediment displacement occurs.

Atkinson and Taylor (1991) provide a thorough description of the mechanical methods by which fish are able to excavate sediment and create burrows. They also describe a variety of burrow morphologies, including vertical burrows or "shafts," horizontal burrows ("tunnels" or "galleries"), and "burrow systems" comprised of a complex burrow with interconnections. Although they provide no quantitative data concerning amount of sediment moved or bioturbation rates, they note that burrow "life" is usually shorter in the field than observed in laboratory simulations of burrows, so that lab estimates of bioturbation rates associated with burrowing fish probably underestimate reality.

Although Atkinson and Taylor (1991) noted a paucity of data on bioturbation and erosion rates associated with burrowing, limited data do exist, some of which has been cited above. Colin (1973) described the morphometry of burrows of the yellowhead jawfish (*Opistognathus aurifrons*) in the vicinity of North Bimini, Bahamas. Burrow casts there ranged in depth from 11 to 22 cm, and the maximum size of burrow chambers was 24 cm long, 23 cm wide, and 6 cm high. Colin also reported on mounds of coral rubble constructed by sand tilefish (*Malacanthus plumieri*).

Additional quantitative burrow data have been reported from Lake Superior by Boyer et al. (1990). They reported on the deep lake-floor sediment mixing produced by burbot (*Lota lota*), which produce deep U- to V-shaped trench burrows with dimensions of 3–5 m long, 15 cm wide, and 10–25 cm deep. Such an individual burrow would displace 0.045–0.188 m³ of sediment. Boyer et al. (1990) also reported average sedimentation rates in Lake Superior and estimated that, given the depth of burrows and associated bioturbation, the last three thousand years of the postglacial sedimentary record may be locally mixed by burbot.

Walls of submarine canyons such as those off the northeastern coast of the United States, carved by fluvial action during the low-sea conditions of the Pleistocene, are susceptible as well to the effects of burrowing fish (Dillon and Zimmerman 1970; Twichell et al. 1985). Block and Corsair canyons, incised into the New England continental shelf, are currently being eroded by the combined effects of fish burrowing and sediment ejection, and subsequent slumping of canyon sidewalls (Dillon and Zimmerman 1970). The combined process also supplies silty suspended sediments to weak currents, allowing for sediment dispersal toward the mouths of the canyons.

Perhaps the most remarkable account of topographic modification by fish comes from an 800-km² area around the head of Hudson Canyon off the eastern coast of the United States. Tilefish and associated crustaceans have created a rough seafloor topography with a local relief of 1–10 m through their burrowing activities (Twichell et al. 1985). Average tilefish burrows there are 1.6 m in diameter and 1.7 m in depth, with an average density of 2,500 per square kilometer! No large mounds of sediment occur around the burrows, suggesting that the displaced sediment is subsequently swept away by bottom currents. Using an average-sized burrow of 1.0 m³ and the aforementioned average burrow density, 2,500 m³ of sediment per square kilometer has been removed by tilefish and their associates (Twichell et al. 1985, p. 717). When extrapolated to the 800-km² study area, this accounts for the excavation of 2.0×10^6 m³ of sediment! Because this rough topography has been produced on seafloor inundated only since the end of the Pleistocene, average erosion rates necessary to produce such a spectacular local relief were calculated. Twichell et al. (1985) determined that an erosion rate of 13 cm per thousand years (or 130 B, where 1 B[ubnoff] = 1.0 mm of erosion per 1,000 yr = 1.0 m^3 km^{-2} yr^{-1}) would be necessary to remove an average depth of 1.7 m of sediment if sea transgression began thirteen thousand years ago; an even higher rate (deemed more likely by the authors) of 18 cm 1,000 yr^{-1} (180 B) is required if erosion started about nine thousand years ago.

The geomorphic role of amphibians

Amphibians as a group are probably the least important geomorphic agent among the ectothermic vertebrates. Only a few examples of geomorphic accomplishments by amphibians are known. The nest-building frog *Leptodactylus marmoratus* digs a hole in the ground in which eggs are laid, after which the hole is then covered again with earth (Hansell 1984). Males of the frog species *Hyla faber*, native to Brazil and Argentina, prepare small pools in the shallow parts of streams for the protection of a batch of eggs. The pool is dredged out by the kicking action of the frogs, which then use the excavated material to build mud ramparts around the pool. These pools are approximately 30 cm in diameter, and the ramparts rise 6–10 cm above the water (Von Frisch 1983; Hansell 1984).

Another example of geomorphic accomplishments by amphibians is more unusual. The male African bullfrog (*Pyxicephalus adspersus*) constructs channels between larger bodies of water and smaller adjacent peripheral ponds containing tadpoles. These channels allow the tadpoles to move into the main water body (Kok, DuPreez, and Channing 1989). They are constructed by using the hind legs like excavating spades. Kok et al. (1989) report that one such channel was 3.2 m long, 150 mm wide, and 20–50 mm deep. An African bullfrog can, therefore, move ~0.01–0.025 m^3 of sediment during channel construction.

Breckenridge and Tester (1961) and Ross, Tester, and Breckenridge (1968) described Mima-type mounds in northwestern Minnesota (see Chapter 7) attributed to the geomorphic work of animals, including Manitoba toads (*Bufo hemiophrys*). The toads apparently utilize the mounds for winter hibernation (Breckenridge and Tester 1961). On one large mound, a total of 3,276 toads displaced a soil volume of 85 ft^3 yr^{-1}. Much of this displaced soil is subsequently bioturbated and mixed back into the mound by the burrowing work of badgers and gophers. The specific long-term accomplishments of the toads are, therefore, open to interpretation.

Other estivation burrows and winter hibernacula of amphibians have been described by Campbell (1970) and Pinder, Storey, and Ultsch (1992).

The latter thoroughly described the diversity of burrow microenvironments, but provided only limited quantitative data as to the amount of sediment displaced by burrowing amphibians. Most burrows, whether in well-aerated soil or in mud either above ground or beneath water, are less than a meter in depth and have only transitory geomorphic influence.

The geomorphic role of reptiles

Hansell (1984) states that building talent in the reptiles is almost absent, but this statement discounts the digging of burrows, referring only to construc-tional features above ground. The reptiles as a group have more influence on surface morphology than do fish or amphibians. This should not be surpris-ing, given that reptiles lay their eggs on dry land, unlike either amphibians or fish. Modern reptiles are also considerably larger than most amphibians, the giant salamander not withstanding, and thus have more sheer bulk and strength for performing geomorphic alterations of the surface.

Among the reptiles, several types of geomorphic work can be identified, and the categories are very similar to those accomplished by fish and am-phibians. These include nesting and associated digging, and burrowing for daily shelter and for seasonal inactivity. Lithophagy, the eating of stones for use as gastroliths in the digestive system, and direct erosion by grazing also accomplish minor amounts of geomorphic work.

Direct erosion by grazing

Viles (1988b) has described the minor erosional effects of a population of giant tortoises (*Geochelone gigantea*) on Aldabra Atoll. Grazing by the tor-toises there produces distinctive grooves, typically 1–5 mm wide and up to 1 m long. These grooves are often found in groups of four or five, and spa-tially are concentrated around drinking holes. Viles also noted minor abra-sion of rock surfaces associated with contact with the tortoises' shells, lead-ing to minor rounding of small rock pinnacles.

Lithophagy

The literature on lithophagy in reptiles, with emphasis on the crocodilian family, has recently been reviewed by Fitch-Snyder and Lance (1993). Reptiles ingest rocks to aid digestive and/or hydrostatic functions (Brazaitis 1969; Sokol 1971). Alligators and caimans use their snouts to loosen gravel, and then pick up rock fragments in their mouths and subsequently swallow them. Fitch-Snyder and Lance (1993) reported a preference by alligators for smoother-edged pebbles than for rough-edge dolomitic rocks, which were picked up but rejected. Alligators, and presumably other crocodilians, are therefore agents of particle-shape sorting! Other reports of lithophagy among the crocodilians include those of Kennedy and Brockman (1965), Brazaitis (1969), and Peaker (1969), and Johnson (1993) reported the ingestion of gastroliths by tortoises. Brazaitis (1969), summarizing previous research, also presented new evidence that ~0.5–1.0% of a crocodilian's body weight is a result of gastroliths in the stomach. If the weight of all crocodilians in the world were known, one could take 1% of that figure to determine the worldwide weight of rock material eroded by crocodilian lithophagy!

Burrowing

Reptile burrows are familiar to anyone who has ever been warned not to put a stick down a snakehole. Most reptiles do not regularly dig burrows for daily occupation (Burger and Gochfeld 1991), although many reptiles use burrows dug by mammals. Nevertheless, some lizards and snakes do excavate their own burrows for nests or for year-round use (Carpenter 1982; Rand and Dugan 1983).

Little quantitative information exists as to the erosional efficacy of reptiles involved in digging and burrow creation. Carpenter (1982) noted that an excavating bullsnake can move as much as 3,400 cm^3 of sediment per hectare. Mora (1989) mapped and described complexes of communal nesting burrows shared by green iguanas (*Iguana iguana*) and ctenosaurs (also known as black iguanas, *Ctenosaura similis*) in Costa Rica. Three simple nest systems revealed 11, 20, and 15 m of tunnels in 10, 13, and 7.5 m^2, respectively. Mora also described two complex systems examined via excavation: a 48-m^2 plot contained 145 m of tunnels, and a 68-m^2 plot had fully 200 m of communal tunnels.

Burger and Gochfeld (1991) also described burrows dug by black iguanas in Costa Rica. Conspicuous piles of dirt fan out from burrow entrances, which were typically dug on slopes of >15°. Average heights, widths, and depths (in cm) for 57 burrows were 10.8, 16.3, and 79.9, respectively. I extrapolated their data on burrow size to a 1-km^2 plot, and calculated that a not-insubstantial 99.68 m^3 km^{-2} of sediment were displaced by black iguanas.

Gopher tortoises (*Gopherus polyphemus*) inhabit well-drained sandy soils in six states on the Coastal Plain of the southeastern United States (Garner and Landers 1981; Kushlan and Mazzotti 1984; Lips 1991; Witz, Wilson,

and Palmer 1991; McCoy, Mushinsky, and Wilson 1993), with its population core in southern Alabama and Georgia and in northern and central
Florida. The tortoises excavate deep burrows for shelter from extreme climatic conditions and predators. Mounds of sandy soil are deposited at the
mouths of the burrows (Kaczor and Hartnett 1990); these soil mounds may
serve as sites for egg deposition by gopher tortoises (Martin 1989). The burrowing and soil-mound deposition create distinct patches that are distinctly
different in edaphic and vegetative characteristics than the surrounding environment (Garner and Landers 1981; Kaczor and Hartnett 1990). As rains
wash mound materials back into burrows (Toland 1991), significant faunalturbation results.

The size of gopher tortoise burrows has been commented upon in several
works. Lips (1991) stated that their burrows averaged 4.6 m long and 2 m
deep. Martin (1989) provided base × height dimensions of 36 × 14 cm, and
30 × 16 cm for two active burrows. I used a mean of Martin's base and
height data, multiplied by Lips's (1991) 4.6-m average length to calculate
an average volume of 2.13 m^3 of sediment removed per gopher tortoise burrow. Kaczor and Hartnett (1990) reported a density of forty-six gopher tortoise mounds and associated burrows per hectare in a study site in central
Florida, whereas Kushlan and Mazzotti (1984) found a density per hectare
of 18.3 burrows at the southern end of the gopher tortoise's range on Cape
Sable, Florida. Using Kaczor and Hartnett's (1990) burrow density and my
calculated average volume, about 98 m^3 ha^{-1} (46 ha^{-1} × 2.13 m^3) of sediment is excavated by gopher tortoises. Kushlan and Mazzotti's (1984) density data, on the other hand, yield a figure of ~39 m^3 ha^{-1}. Assuming that
these burrow density values are representative, then, one can generalize that
gopher tortoises faunalturbate and excavate ~40–100 m^3 of sediment per
hectare.

Most studies of gopher tortoise burrows revealed no tendency for orientation of the burrows. Recently, however, McCoy et al. (1993) conducted a
landscape-scale determination of burrow orientations for more than 3,500
burrows in a total of thirty-seven gopher tortoise populations in Florida.
Their data revealed that burrows tend more often to be oriented in the primary (N, E, S, W) compass directions than in the secondary (NE, SE, SW,
NW) ones. They attributed this finding to the orientation of topographic
features that are themselves oriented in the primary compass directions.

Nest building by reptiles involves little actual geomorphic excavation and
transport, because while many reptiles do indeed excavate a pit in which to
deposit eggs, the pit is subsequently infilled by the same sediment after egg
deposition occurs. (Propper et al. [1991] provide a step-by-step description

of the process as carried out by the American chameleon, *Anolis carolinen-sis.*) Added to the overall geomorphic insignificance of nest building and egg deposition is the fact that many lizards and turtles carry out the action in soft beach sediments (Burger 1993), where eolian and winter wave action can quickly erase the effects of the reptiles. Mud turtles (*Kinosternon fla-vescens* and *Kinosternon subrubrum*) may bury themselves in mud along with the eggs (Burke, Whitfield Gibbons, and Green 1994), but the lasting geomorphic significance of this action is open to debate.

The major exception to the geomorphic insignificance attached to most members of the reptile class is the excavational, faunalturbational, and tram-pling effects produced by members of the crocodilian family. Crocodilians, including alligators, crocodiles, gavials, and caimans, are the largest mem-bers of the reptile family and are found in tropical environments around the world (Table 3.1). They have four primary activities that influence the geo-morphology of their habitat: nesting, wallow digging, denning, and tram-pling. These activities were recently summarized by Wheeler (1991), from which the following accounts are excerpted.

Crocodilian nesting

The nests of crocodilians are of two types, mounds and holes (Greer 1970). Mound nests consist of a pile of vegetation intermixed with soil material, whereas hole nests are excavated into the ground (Campbell 1972). Table 3.2 lists the type of nest and average amount of displaced sediment for a variety of crocodilians. Kofron (1989) described three general shapes of nest cavities, depending on substrate conditions. Bowl-shaped nests were dug in loose sand, funnel-shaped nests in loose sand underlain by firm substrate, and downward-slanting nests with overhanging roofs in firm soil. After egg deposition, a layer of sediment 10–30 cm thick is redeposited over the eggs.

The majority of crocodilians are mound-nest builders (Wheeler 1991). The mound nests of the American alligator (*Alligator mississippiensis*) may be examined as typical. A mound nest is dome-shaped, about 50 cm high, and with long- and short-axis diameters of 1.6 and 1.5 m, respectively (Goodwin and Marion 1978). Vegetation is flattened in a 6–8-m-diameter area, then is torn or bitten off and placed in the center (Joanen and Mc-Nease 1989). Mouthfuls of mud and swamp grass are added to the pile, and the mound is compacted by the female crawling over and around the nest. The direct erosional action of sediment excavation combines with the indi-rect effects of vegetation removal and trampling to make crocodilian nest building a locally significant geomorphic process.

Table 3.1. *Crocodilian size and geographic range*

Species	Maximum size (m)	Range
Gavial gangeticus (Indian gavial)	9.1	Northern India
Tomistoma schlegeli (Malayan gavial)	4.6	Borneo & Sumatra
Crocodylus cataphractus (sharp-nosed crocodile)	3.7	West Africa
C. johnsoni (Australian crocodile)	2.4	Australia
C. intermedius (Orinoco crocodile)	3.7	Venezuela
C. acutus (American crocodile)	4.3	Florida; Mexico; Central & South America
C. siamensis (Siamese crocodile)	2.1	Thailand, Java
C. niloticus (Nile crocodile)	5.5	Africa (in general)
C. porosus (saltwater crocodile)	6.1	India & Malaysia
C. robustus (Madagascar crocodile)	9.1	Madagascar
C. rhombifer (Cuban crocodile)	2.1	Cuba
C. moreletti (Guatemalan crocodile)	2.1	Guatemala, Honduras
C. palustris (swamp crocodile)	4.9	India & Malaysia
Osteolaemus tetraspis (broad-nosed crocodile)	1.8	West Africa
Caiman trigonotus (rough-backed caiman)	1.8	Upper Amazon
C. sclerops (spectacled caiman)	2.4	Central & South America
C. palpebrosus (banded caiman)	2.4	Tropical South America
C. latirostris (round-nosed caiman)	2.4	Tropical South America
C. niger (black caiman)	6.1	Tropical South America
Alligator mississippiensis (American alligator)	4.9	Southeastern USA
A. sinensis (Chinese alligator)	1.8	China

Sources: Data from Ditmars (1936) and Wheeler (1991).

Table 3.2. *Crocodilian nest data*

Species	Nest type	Approx. volume (m^3)
Alligator mississippiensis	Mound	0.15
A. sinensis	Mound	0.03
Caiman crocodilus crocodilus	Mound	0.05
C. crocodilus yacare	Mound	0.08
C. latirostris	Mound	0.09
Osteolaemus tetraspis tetraspis	Mound	0.02
Crocodylus acutus	Mound/hole	0.15
C. cataphractus	Mound of sand	0.03–0.06
C. johnsoni	Hole	0.0004
C. moreletti	Mound	0.80
C. niloticus	Hole	0.01
C. palustris	Hole in sand (or mound?)	0.004
C. porosus	Mound	0.13

Sources: Data from Ferguson (1985) and Wheeler (1991).

Wallows

Crocodilians also construct and maintain wallows, or "gator holes," so-called because the wallows of the American alligator are the best known. Gator holes range in size from a few meters to small lakes. They are simply depressions found in swamps that are kept free of vegetation by the alligator. Wallows dug by alligators can be up to 2 m deep and 3 m or more in diameter, and mud is piled around the perimeter of the hole. The wallows serve as important microhabitats for other wildlife and plants around their edges, serving to create patch environments that would otherwise not exist; if not for continuous alligator activity, the gator holes would fill within a short time (Wheeler 1991).

Denning

Crocodilian dens are excavated at the base of large ponds, in the sides of river banks, and beneath tree roots such as in mangrove swamps (Wheeler 1991). Dens are constructed primarily for periods of winter inactivity, but may also be used for periods during nesting and postnesting (Joanen and McNease 1989). An American alligator can dig a den big enough to fit its entire body within a few days, but continues to enlarge the tunnel for many years. Tunnels over 20 m in length have been recorded (Wheeler 1991), diameters are roughly proportional to the size of the crocodilian, and multiple branches and outlets to the surface exist.

Trampling and trails

Alligator migration trails are created by repetitive movement over the same path of soft mud or sand (Joanen and McNease 1989; Wheeler 1991). Over time, troughs are formed that can be 15 cm deep and 60 cm wide. A 30-m-long trail would therefore erode or compact 2.7 m^3 of sediment. Such trails may also have the indirect effect of exacerbating fluvial erosion by serving as stream bypasses.

Conclusions

The ectothermic animals collectively perform relatively minor amounts of geomorphic work. In local cases, such as fish in submarine canyons or in areas of concentrated tortoise, iguana, or crocodilian populations, fish and reptiles may produce more sediment erosion and transportation than would be expected at first glance. Amphibians are probably the least significant geomorphic group in the entire animal kingdom, but even they produce local landforms and accomplish sediment movement. Oh, to have been a geomorphologist examining sediment movement in the Permian period or the Mesozoic era!

4

Birds as agents of erosion, transportation, and deposition

Introduction

Birds are relatively lightweight animals, and many species spend much of their life cycles in the air, in aboveground nests, and/or in the water. Nevertheless, distinct geomorphic effects can be attributed to some birds. Some are relatively minor geomorphic curiosities, whereas others induce widespread alteration of the landscape in which the birds live. This chapter categorizes the geomorphic effects of birds and examines some specific cases by which these effects are produced.

The geomorphic effects of birds include internal clast transport as gastroliths; geophagy; external clast transport for use as tools ("bioports," *sensu* Johnson 1993); clast and mud transport for use in nests; mound building and surface scraping; vegetative removal; and burrowing and nest-cavity excavation. Each of these processes is examined below.

Internal clast transport as gastroliths

It is well known that birds ingest sediment to act as grinding tools in the gizzard, in a fashion similar to that employed by reptiles (see Chapter 3). An almost complete void of quantitative data exists, however, as to the amount of erosion and transport involved in avian lithophagy. Milton et al. (1994) reported on the stone contents of ten ostrich gizzards taken in southern Africa. Stone mass in the adults averaged 0.646 ± 0.266 kg, and in subadults averaged 0.444 ± 0.266 kg. Johnson (1993, p. 74) provided data on the number, size, and weight of stones from five ostrich gizzards. These included 5,000 stones totaling 0.855 kg, 2,856 stones totaling 0.707 kg, 9,128 at 0.991 kg, 3,108 weighing 0.433 kg, and 435 at 0.117 kg. Long axes of the largest gastroliths from each of the five birds measured 16.5, 18.1, 19.0,

19.1, and 21.7 mm, respectively. Gastroliths up to 2.5 cm long have also been reported in the literature, as described by Johnson (1993). He further noted that numerous large birds can carry thousands of stones in their gizzards, and mentioned – in addition to the ostrich, the emu, the cassowaries, the rheas, and recently extinct birds such as the moas of New Zealand – the elephant birds of Madagascar and the mihirungs of Australia. The gastroliths carried by large birds are, according to Johnson, undoubtedly incorporated as residual elements in many soil stone lines.

Geophagy

Geophagy, or earth consumption, differs from lithophagy in that fine-grained material is ingested presumably in order to rectify mineral deficiencies in bird diets (Jones and Hanson 1985) or to counteract (i.e., detoxify) the effects of ingested toxic food sources (Munn 1994). Jones and Hanson (1985) point out the almost complete dearth of research on bird geophagy, and cited only anecdotal accounts of earth consumption by birds. Recently, Beyer, Connor, and Gerould (1994) noted that sandpipers (*Calidris* spp.) that probe or peck for invertebrates in shallow water or mud consumed sediments at a rate of 7–30% of their diets. They also noted the consumption rates (as a percentage of diet) for Canada geese (*Branta canadensis,* 8%) and wild turkey (*Meleagris gallopavo,* 9%).

Munn's (1994) recent account of parrot and macaw geophagy at a clay-rich river bank in the Peruvian Amazon is one of the few eyewitness accounts of geophagy in birds. Although lacking in quantitative data, the article vividly describes how more than a thousand parrots "squabble over choice perches to grab a beakful of clay, a vital but mysterious part of their diet. More than a dozen parrot species will visit the clay lick throughout the day . . ." (p. 123). The absence of quantitative data on the effects of avian geophagy is clearly an opportunity for future research.

Clast transport for use as tools

Another example of minor geomorphic impact deals with the fact that some birds transport pieces of rock for use in cracking open food sources. The habit of dropping stones onto eggs has been recorded for Egyptian vultures (*Neophron percnopterus*) (Van Lawick-Goodall 1968; Andersson 1989; Thouless, Fanshawe, and Bertram 1989) and fan-tailed ravens (*Corvus rhipidurus*) (Andersson 1989), the latter of which have also been seen to hold a stone in the bill as a distinct tool during hammering. Egyptian vultures carry rocks for use in cracking open ostrich eggs (Bertram 1992) and have been seen to transport such rocks over distances generally up to ~50 m (Van Lawick-Goodall 1968), with unusual cases of up to 200 m reported by Thouless et al. (1989). However, the overall geomorphic results of clast transport for tool use by birds are unquantified although clearly minor.

Transport of materials for use in nest building

More significant, but similarly unquantified, is the erosion and transport of stones and invertebrate shell fragments for use in nests. Cadée (1989) has examined the question of shell-fragment transport by birds, and Hobson (1989) described the use of pebbles in the nest building of double-crested cormorants (*Phalacrocorax auritus*). Pebbles were found in 0.1–6.2% of the nests of the birds examined, and the size of pebbles used in cormorant nests averaged approximately 4 cm (range, 0.5–10 cm). The number of pebbles used, which could provide a crude gauge of the geomorphic effect of this action, was not reported.

The use of pebbles by colonial seabirds such as penguins and Atlantic Alcidae has been reported by several investigators (Nettleship and Birkhead 1985). Harris and Birkhead (1985) noted that the razorbill (*Alca torda*) collects small stones upon which eggs may be laid, and the dovekie auk (*Alle alle*) frequently lays its eggs on a bed of rock pebbles 1–4 cm in diameter. Particularly well known are the pebble nests and guano concentrations of penguins along the coast of Antarctica and on some subantarctic islands. Mitchell (1988) described several important aspects of the effects of extensive nitrogen-rich guano droppings on nutrient cycling in the region, and Heine and Speir (1989) and Baroni and Orombelli (1994) have described pebble concentrations and guano-rich organic soils deriving from Adélie penguins (*Pygoscelis adeliae*). Currently utilized rookeries have widespread low, roughly circular mounds of pebbles and subangular stones that were collected by the penguins for the nest. Nests in high-density locations, such as in the Ross Sea region of Antarctica, are spaced only about 1 m apart (Heine and Speir 1989). Between the nests, the surface is capped with stone-free guano crust. Through time, layer upon layer of guano and stones have been built up, until deposits of pebbles, guano, and buried ornithogenic soils nearly 1 m thick have accumulated (Heine and Speir 1989) over the course of the entire period since Pleistocene deglaciation, ~11–13 thousand years before the present (Baroni and Orombelli 1994). Gravel and stones in the guano-rich horizons are mainly 10–40 mm in diameter, apparently the size of preference for penguins to carry in their beaks easily (Heine and Speir

1989). Literally millions of pebbles have been moved by penguins in such rookeries, but lack of data on the spatial extent, depth, and total numbers of penguin nests and paleoornithogenic soils precludes attempts at quantitative assessment of the overall geomorphic role of penguin-nest construction.

The use of mud as lining or constructional material for nests has been described in general terms by Terres (1980) and Hansell (1984). Among the thousands of species of birds, it is estimated that mud is used in the nests of not more than 5% of bird species, including those that use it only as a minor component as a plastering agent on the interior of a nest. Mitchell (1988) noted that of the more than 700 bird species in Australia, only 10 construct mud nests in trees. A number of species of thrush (Turdidae) and some New World blackbirds (Icteridae) utilize mud in their nest building (Hansell 1984, p. 83). Von Frisch (1983) and Hansell (1984) also provided specific descriptions of nests constructed of mud. These included the nests of the house martin (*Delichon urbica*), the crag martin (*Hirundo rupestris*), the blue swallow (*H. atrocaerulea*), and the oven bird (*Furnarius leucopus*), so-called because it creates a large domed nest with an access tunnel that looks strikingly like a kiln or oven. Details of this constructional process are provided in Von Frisch (1983).

The nests of flamingos (Phoenicopteridae) are constructed entirely of mud, and one of the earliest geomorphic descriptions of these mounds was in Lobeck's (1939) text. He noted that flamingo nest mounds in the Bahamas are formed from soft lime mud at low tide, and range from 1 to 2 ft (~30–60 cm) in height. Some more detailed quantitative measures of the amount of mud used have been provided by Brown and Root (1971). They described the nest of the lesser flamingo (*Phoeniconias minor*) as being a mud column 22–30 cm high and 27 cm across the top, with sides angled 70–80° to the horizontal. The average weight of a nest was ~20 kg. The great flamingo (*P. ruber*) builds mud nests that are even taller and broader, and that weigh an average of 52 kg. In Brown and Root's study area in Tanzania, over one million flamingo nests were constructed in one breeding season, involving over twenty thousand tonnes of soda mud (1971, p. 159)!

Recently, Gauthier and Thomas (1993) described the nest-building habits of cliff swallows (*Hirundo pyrrhonota*) at a site in Sherbrooke, Québec. Swallows "commuted" distances of 50–320 m to garner mud for use in nest construction. Mud sources were not constant, changing almost daily depending on recent precipitation and the consistency of various mud sites (p. 1121). Individual mud pellets used in nest construction averaged 0.36 g in weight, such that a detached nest weighing 652.8 g (mean data for thirty-seven nests) required 1,813 mud pellets for construction; twenty-four semi-

detached nests averaging 602.7 g required 1,674 mud pellets; and row nests
($n = 8$) weighing an average of 573.1 g required 1,592 mud pellets. *If* these
data are representative of cliff swallows worldwide, and *if* an accurate census of breeding pairs were available, then the worldwide geomorphic effect
of nest building by cliff swallows would be calculable. The necessary data
are, of course, unavailable, illustrating the difficulty in quantitatively assessment of the effects of cliff swallows, or any bird for that matter, over
more than a local geographic scale. Burrow-excavating aspects of cliff swallows are addressed in a later section of this chapter.

Mound building

Associated with nest building and geomorphic work are surface scraping
and the creation of incubation mounds. Building of incubation mounds is
practiced by several species of *megapode* (a specialized western Pacific–
Australasian group of birds with large powerful legs and feet) (Troy and
Elgar 1991). Mitchell (1988) noted that eleven species of Australian birds
rake forest litter, forty-four nest in shallow scrapes on the soil, and three
build large incubating mounds of soil and litter. Particular attention has been
given to the work accomplished by the Australian brush turkey (*Alectura
lathami*) and the lyrebird (*Menura novaehollandiae*).

   Male brush turkeys build large incubation mounds by raking together
great quantities of soil and leaf litter with their claws. Troy and Elgar
(1991) reported that a male brush turkey takes an average of thirty-six days,
working 5–7 hr per day raking and moving soil and litter, to build a mound
into which eggs are laid, and that subsequent maintenance of the mound
occupies small amounts of time each day for an additional three to five
months. They estimate that a male brush turkey may move as much as
2,500 kg of moist soil and litter to construct such a mound; unfortunately,
they made no quantitative distinction between the amounts of soil and leaf
litter, making extrapolations of this nature unreliable at present.

   Lyrebirds are moderately large ground-feeding birds found throughout
most of southeastern Australia. Mitchell (1988) provided the most detailed
account of their mound-building behavior, from which the following descriptions are excerpted. In the months before breeding, the male lyrebird
builds as many as eighty or more display mounds. These are built by clearing vegetation and litter from a 1-m-diameter circle and raking up soil to
form a mound 10–15 cm high. In his study area 85 km WNW of Sydney,
New South Wales, Mitchell examined twenty-two lyrebird mounds. These
averaged 1.03 m^2 in area, 12.4 cm high and 4.7 cm thick, and contained an

average of 24 kg of soil. They were circular in plan, with an asymmetric pattern of upslope cut and downslope fill on slopes >4°. Soil excavation could in some cases produce a face up to 23 cm high. The soil around the perimeter of the mounds was disturbed to an average depth of 9 cm. Mitchell (1988, pp. 59–61) reported that lyrebird mounds occupied 1% of his study area, and had a calculated soil turnover rate of 0.4 t ha^{-1} yr^{-1}, leading him to conclude that total soil turnover to a depth of about 10 cm could occur in a thousand years. Mitchell also noted the feeding activities of lyrebirds in this area, which are described in a subsequent section of this chapter.

A different, nonincubational kind of mound created by birds was described by Verbeek and Boasson (1984) from a high cirque in the western Pyrenees of France. Near the top of escarpments and small, rounded hills, "one sees small, mole-hill sized, vegetated hummocks, one or two per knoll" (p. 337). These mounds, or hummocks, ranged in height from 10 to 28 cm, with an average height of 20.7 cm. The soils of the hummocks were considerably more acidic than the surrounding soils, and Verbeek and Boasson attributed the hummocks' development to the collection of bird droppings on the knolls and escarpments. They noted that these topographically elevated locations received two orders of magnitude more bird droppings a day than lower nonmound or nonescarpment sites, and speculated that the high-ground sites were used for nesting sites, feeding stations, and/or lookouts:

[I]n the course of time, territory-owners will have perched in those places where they had maximum visibility of the sky and ground. . . . Because these locations were used more often than others, they also received more droppings along with their beneficial nutrients for plants. One or more of the spots eventually became used more often than others, developed relatively luxuriant vegetation, and once this had occurred, a self-perpetuating system resulted in the rise of the hummock. (Verbeek and Boasson 1984, p. 341)

The mounds therefore owed their existence to collections of guano and the subsequent fertilization effects, encouraging plant growth and the rising of a hummock on topographically elevated positions. Verbeek and Boasson (1984) also cited anecdotal accounts of such low hummocky mounds in the Arctic environment, but no attempts at determining their overall geographic extent or geomorphic influence were made.

Vegetation removal and its geomorphic effects

Although the mound-building behaviors of the Australian brush turkey and lyrebird are somewhat unusual, other species of birds scrape the ground surface clean in order to breed, nest, and lay eggs. Vegetation removal by birds may also be associated with feeding, or with movements across the landscape. These processes are described below. Specific geomorphic actions associated with actual burrow construction and burrow sites are discussed in a subsequent section of this chapter.

Geomorphic effects associated with breeding sites

A classic example of scraping the ground clear of vegetation for courtship, breeding, and egg-laying purposes is that of the African ostrich (*Struthio camelus*) – whose burying of its head in the sand has *never* been actually observed and is merely a longstanding myth (Bertram 1992). Male ostriches make several scrapes in the soil, apparently as an invitation to the female for mating. The process involves scratching of the soil with powerful toes, and subsequent sitting and grinding of the body into the soil (Bertram 1992, p. 51). Eventually, one such scrape is accepted by a female, and becomes a nest when she lays the first egg in it. Although Bertram's (1992) purpose was to describe mating and egg-laying behavior rather than the geomorphic effects of scraping and nesting, photographs included in his book suggest that a typical scrape/nest site ranges up to 2 m in diameter, with a depth of disturbance of perhaps 5 cm. Vegetation is completely removed from the site.

The effects of large colonies of seabirds breeding in spatially concentrated locations result in extensive removal of surface vegetative cover. Mitchell (1988) provides a succinct review of this topic, with special emphasis on Australian bird populations. Subterranean burrow-nesting birds particularly impact the surface vegetative cover by inducing instability and collapse, but surface-breeding species may also produce geomorphic impacts. Joly, Frenot, and Vernon (1987) described surface trampling and vegetative destruction associated with the prolonged nesting period of breeding wandering albatrosses (*Diomedea exulans*) on subantarctic Ile de la Possession in the Indian Ocean. They also noted the modification of soil cover on the island, associated with large amounts of accumulated organic deposits (droppings, feathers, and regurgitations) from these birds.

Hall and Williams (1981) also described concentrations of surface-breeding albatrosses (*Diomedea* and *Phoebetria* spp.), in their case from a study on subantarctic Marion Island. There, steep slopes (35–60°) may be maintained beyond the angle of repose by the binding action of ground vegetation. Albatrosses remove the vegetation for nest material, undermining the stability of the slopes. When saturated, and upon the shock impact of landing by the birds, the shear strength of the slope material is overcome and localized slumping ensues. Hall and Williams (1981, p. 19) described areas on the cliffs of Macaroni Bay, Marion Island, where "units of vegetation and soil in the region of 0.4 × 0.3 × 0.1 m were seen to be removed." They did not describe how many of these 0.01-m^3 slumps existed, but also noted that slumping is common on relatively steep (16–35°) slopes where burrowing by penguins (described in a later section of this chapter) has removed surface vegetation and undermined the surface.

Breeding densities of penguins in the subantarctic and Antarctic regions can be very high, and examples are summarized in Müller-Schwarze (1984). The result of these high densities – Hall and Williams (1981) cite densities of 2.3 and 4.5 pairs per cubic meter for king (*Aptenodytes patagonicus*) and macaroni (*Eudyptes chrysolophus*) penguins, respectively – is complete vegetative destruction and removal in association with widespread trampling and manuring. Removal of the surface mat of vegetation in these delicate environments allows the penguins subsequently to claw and remove the underlying soft, wet, peaty subsoil (Hall and Williams 1981). At two colonies on Marion Island, a peat cover of up to 4-m thickness has been totally removed in this fashion from areas of 82,000 and 110,000 m^2. Hall and Williams (1981) assumed an average peat thickness of 1.5 m over the two areas, resulting in a total peat removal from the two colonies of 2,900 m^3. They also cited another example where severe erosion associated with king penguins covered an area of 630,000 m^2.

Trampling and grooving of bedrock

Movements of large colonies of flightless seabirds to and from the sea and inland breeding sites establishes natural walkways where vegetation is rapidly destroyed and erosion is concentrated. Gentoo penguins (*Pygoscelis papua*) return from the sea to their on-land breeding colony every evening throughout the breeding season on Marion Island (Hall and Williams 1981). During these periods of daily passage, the gentoo penguins wear deep tracks through the vegetation and peat surrounding their landing beaches. These tracks expose the surface material to eolian removal, and may also form local drainage lines that result in exacerbated fluvial erosion.

Grooving and polishing of bedrock caused by the direct action of bird claws has been described from several Antarctic and subantarctic locations (Hall and Williams 1981; Splettstoesser 1985; Cooper and Brown 1990), and is also typically associated with passage to and from inland breeding sites. Splettstoesser (1985) briefly described 2–5-cm-deep grooves on near-vertical faces of rock along the shoreline of New Island, Falkland Islands, and attributed their formation to rockhopper penguins (*Eudyptes crestatus*). He described the method of formation as follows: "Over the many, but unknown years that these birds have traveled back to their nests across these rocks their three-clawed, webbed feet have incised numerous grooves, or striations, in many locations on the cliff faces and boulders that are strewn along the beach" (1985, p. 108). He observed the penguin striations primarily on sandstone, but also noted their presence on dolerite dikes.

Hall and Williams (1981) provided some of the most detailed descriptions of the grooving and polishing abilities of penguins. Their examination of penguins on Marion Island, home to about 3.4 million penguins, revealed that king and macaroni penguins are the primary agents of erosion in their passages from onshore breeding sites to the sea. Along one major routeway, about two hundred thousand pairs of macaroni penguins must pass from a narrow landing beach up a steep funneling valley only about 30 m wide. All vegetation and soil has been removed through this valley, and the gray lava bedrock has been grooved and polished "to a mirror-smooth finish" (Hall and Williams 1981, p. 19). The grooves in this walkway are up to 150 mm long, 10 mm deep, and 6 mm wide. Grooving of bedrock by rockhopper penguins (*Eudyptes chrysocome*) is apparently absent on Marion Island, but has been reported from Gough Island and the Falkland Islands (Hall and Williams 1981).

Surface scraping and feeding effects

Surface scraping and clearance of vegetation in association with ground feeding is widespread among several species of birds, but the actual geomorphic impacts of this process is poorly quantified. Most attention has been focused on the feeding scrapes of Australian lyrebirds, geese, and shorebirds.

Mitchell (1988) provided a quantitative description of the geomorphic effects of lyrebird feeding activities in southeastern Australia, as well as excellent visual evidence of the efficacy with which these birds excavate the surface in their search for food (see his figs. 2.3–2.5). Feeding by lyrebirds in Mitchell's study was conducted almost exclusively on slopes, and involved disturbance and shifting of litter and the soil to depths of 5–10 cm (1988, p. 61). Feeding on slopes >8° is accomplished by the birds facing uphill and kicking excavated debris about 1 m downslope. Rocks weighing up to 2 kg were frequently shifted downhill in such a fashion. A concentration of feeding sites produced low cliffs of soil 12–15 cm high, which were advanced upslope via the kicking of debris downslope. Feeding concentration was also noted around trees, logs, and rocks, in some cases to the point where the rocks or logs would collapse downhill. The net result of all these actions was "a series of discontinuous terraces and plumes the ultimate size of which varied with slope angle. Plumes of debris accumulated against shrubs, logs and rocks and were often cantilevered over the slope" (p. 62). Mitchell calculated that lyrebird feeding disturbed the soil at a rate of 63 t ha^{-1} yr^{-1}, of which 45 t was mineral soil. From this, he concluded that the soil on 1 ha of the slope could be overturned to a depth of 8 cm in thirteen years, within the lifespan of a single bird! He also noted that as the process advanced, new areas were worked before old workings were reutilized and reactivated.

The direct and indirect geomorphic effects of large populations of feeding geese have been described from several locations in salt-marsh and intertidal environments in eastern and northern Canada (Jefferies, Jensen, and Abraham 1979; Dionne 1985; Jefferies 1988), and have also been briefly described from wintering sites in the U.S. southern Gulf States (summarized in Jefferies et al. [1979] and Dionne [1985]). The work of Jefferies and associates (1979; Jefferies 1988) focused on the effects of between ten and thirty thousand lesser snow geese (*Anser caerulescens*) in salt marshes along the Manitoba shore of Hudson Bay, Canada. In spring, before the onset of aboveground plant growth, but following snowmelt, geese grub for the roots and rhizomes of *Carex* and *Puccinellia,* two major components of the salt-marsh vegetative complex. This grubbing creates widespread patches of bare sediment where geese have literally stripped away the shallow turf. The effects are particularly exacerbated around the turf fringes of small ponds occupying abandoned drainage channels. A goose can strip an area of 1 m^2 of turf fringe in about an hour, and this grubbing is combined with excessive trampling, resulting in the collapse of the banks of the pools. In such manner, the pond margins are enlarged, and a small-scale terrace is created. A series of terraces may be produced as the zone of vegetation that has been stripped extends back into surrounding areas supporting mounds of willows. Excellent photographs of these grubbing-produced pools and terraces are available in Jefferies (1988).

Quantitative data concerning the grubbing effects of lesser snow geese is limited, but Jefferies (1988) noted that it is patchy in both time and space. The scale of the patches on the intertidal flats in his study area at La Perouse Bay, Manitoba, varied from the size of individual beak marks to areas of bare sediment 300 × 300 m in extent. The overall result is the creation of "peat barrens."

Jefferies et al. (1979) also noted the effects of summertime grazing by the lesser snow geese, where grubbing is less significant but trampling associated with the grazing produces small bare areas that accumulate water. These pools may be subsequently enlarged by freeze–thaw action of ice, illustrating the secondary geomorphic influence of the geese.

Dionne (1985) reported similar grubbing and geomorphic impacts associated with greater snow geese (*Anser caerulescens atlanticus*) in the Montmagny tidal marsh along the St. Lawrence River estuary in Québec. There, geese trample and destroy vegetation in a fashion similar to that described above. They also dig thousands of small holes, 6–12 cm in depth and 10–25 cm or more in diameter, in search of desired roots. This creation of several thousand holes occurs at each low tide in May, and September–October, when geese are present. The high density of geese is such that large areas are pitted and covered by coalescing goose holes; excellent illustrations showing these features are found in Dionne's paper. The sediment expelled from the holes dug by geese is left at the surface, where it is subsequently resuspended and carried away by tidal currents. Dionne estimated that this combined action of goose pitting and tidal activity lowers the tidal-marsh surface an average of 8–10 cm annually. Geese in his study area also cause backwearing of low bluffs along the tidal-marsh margin, in a fashion similar to the terrace creations described by Jefferies et al. (1979), and also destroy peat blocks by their grubbing activities.

The bioturbational activities of shorebirds have been examined by Cadée (1990) on a tidal flat of the Dutch Wadden Sea. He described characteristic feeding troughs up to 3 m long, 15 cm wide, and 3 cm deep made by gulls (*Larus ridibundus*), and feeding craters made by shelducks (*Tadorna tadorna*) that are about 10 cm deep and up to 60 cm in diameter. Cadée estimated that approximately 30% of his study area was reworked annually by gulls and about 15% by shelducks. The annual sediment-reworking rate equaled a sediment layer 2.5 cm thick, an order of magnitude smaller than that reworked by polychaete and bivalve deposit feeders (Cadée 1976, 1979).

On the opposite end of the spectrum of geomorphic influence, Daborn et al. (1993) recently suggested that the feeding habits of migratory shorebirds such as the semipalmated sandpiper (Fig. 4.1) (*Calidris pusilla* L.) may enhance the geomorphic stability of shoreline sediments through their interactions with amphipods. Amphipods, particularly *Corophium volutator,* graze on benthic diatoms in the tidal-flat sediments of the Bay of Fundy. The diatoms produce cohesion-developing carbohydrates that enhance sediment cohesion and stability. When the diatoms are removed by amphipods, how-

Figure 4.1. Tidal zone and pool frequented by seabirds along a barrier island; Hunting Island State Park, South Carolina. The geomorphic influence of such seabirds is controversial.

ever, the production of cohesion-inducing carbohydrates is reduced, fine sediments winnow away, and bedforms characteristics of noncohesive sediments appear. The arrival of migratory shorebirds results in substantial decline of the cohesion-inhibiting amphipods; it has been estimated that one sandpiper may feed on more than ten thousand *Corophium* per day. The decline in the number of amphipods, attributable to shorebird arrival, thus produces secondary consequences that reflect greater production by benthic diatoms, which in turn result in cohesive strength (i.e., decreased erodibility) of shoreline sediments. The work of Daborn et al. (1993) both illustrates the complexity of interactions leading to, or inhibiting, geomorphic effects of animals, and suggests that much additional research is necessary to assess, say, the geomorphic effects of other widespread shorebirds such as gulls.

Burrowing and nest-cavity excavations

Ornithological literature indicates that burrowing/nest-cavity excavating is uncommon among birds (Terres 1980). Burrowing is most common among certain groups of colonial seabirds, and is the normal mode of nesting for petrels, shearwaters, storm petrels, diving petrels, and some auks and penguins (Furness 1991). Among terrestrial species, burrowing is more uncommon. Those terrestrial birds that do excavate burrows or nest cavities include bee eaters (*Merops* spp.), kingfishers of the subfamily Alcedininae, and bank swallows (*Riparia riparia,* also known as sand martins) (Furness 1991). The northern rough-winged swallow nests in bank cavities, but probably does not dig its own burrows, instead occupying and simply enlarging preexisting cavities produced by bank swallows or other animals (Gaunt 1965; John 1991). Similarly, other birds that occupy ground burrows (e.g., kakapos, *Strigops habroptilus;* shelducks, *Tadorna* spp.; mountain plovers, *Charadrius montanus;* killdeer, *C. vociferus;* horned larks, *Eremophila alpestris;* and burrowing owls, *Athene cunicularia*) probably normally occupy preexisting burrows dug by animals such as rabbits, ground squirrels, yellow-bellied marmots, badgers, and prairie dogs (Coulombe 1971; Knowles, Stoner, and Gieb 1982; Rich 1984, 1986; MacCracken, Ruesk, and Hansen 1985; Green and Anthony 1989; Furness 1991), although burrowing owls create at least minor modifications to host burrows by digging (Thomsen 1971). In the following discussion, I first examine burrowing and nest-cavity excavations by terrestrial species before examining the more extensively studied colonial seabirds.

Burrows and cavity excavations of terrestrial birds

Bank swallows are well adapted for digging, and excavate nest cavities in banks where at least a 3-m vertical face can be maintained (John 1991). Soil characteristics also dictate where excavations occur: High sand content, low organic content, and sufficient stability are characteristic of sites occupied by bank swallows (John 1991). The burrows are dug not by pecking or drilling, but by lateral slashing at the soil with the bill (Gaunt 1965). Several such slashed-out burrows are illustrated in Gaunt's (1965) examination of bank swallows in Kansas.

Gaunt's (1965) work is one of the few that provide quantitative data on the depth of excavation of bank swallow nest cavities. He dug out twenty-eight nest tunnels and found them to range in depth from 38 to 122 cm, with a mean depth of 70 cm. He also inserted a rod into 122 tunnels, revealing a range of depths of 38–130 cm, with a mean of 69 cm – that is, virtually the same data as provided by the excavation technique. Unfortunately, no diameter data were provided, nor is it known how many such nest cavities exist in North America, making quantitative extrapolations meaningless. Gaunt also noted, however, that minor slumping of banks heavily excavated by swallows is common, indicating that bank swallow nest-cavity excavation has both primary, direct geomorphic effects as well as secondary, indirect influences.

Geomorphic effects of burrowing by colonial seabirds

Colonial seabirds occupy insular and coastal locations in environments ranging from near-tropical to Antarctic. Two sets of factors are of primary importance to nesting seabirds: proximity to feeding sites and security of burrow sites (Birkhead and Harris 1985; Kaiser and Forbes 1992). Burrows offer several advantages for breeding seabirds: security from predators, reduced requirements for nest materials, enhanced thermal stability in environments that may experience temperature extremes (both hot and cold), and reduced heat loss by chicks (LaCock 1988; Kaiser and Forbes 1992). Optimum conditions for burrows for many seabirds include soils of ≥30 cm depth, with a large amount of fine-grained particles and little gravel or sand, and sloping ground rather than flat surfaces (Nettleship 1972; LaCock 1988; Stokes and Boersma 1991; Kaiser and Forbes 1992). Birkhead and Harris (1985) and Harris and Birkhead (1985) caution, however, of "making too much" of the necessity for sloping ground, at least in the case of Atlantic

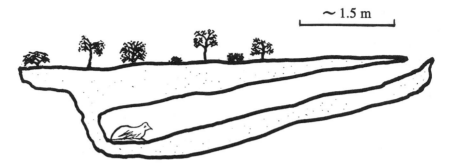

Figure 4.2 Sketch of wedge-tailed shearwater (*Puffinus pacificus*) burrow. Adapted from Dyer and Hill (1990, 1991).

puffins (*Fratercula arctica*): They cite several examples where flat ground serves equally well in providing sites for burrows. The wedge-tailed shearwater (*Puffinus pacificus*) by necessity also burrows in flat surfaces comprised of coral sands on sandy cay islands off the central Queensland, Australia, coast (Hill and Barnes 1989; Dyer and Hill 1990, 1991, 1992).

Burrowing is the normal mode of nesting for several species in the Procellariiformes (petrels, shearwaters, storm petrels, diving petrels), Lariformes (auks), and Sphenisciformes (penguins) orders (Furness 1991). Hill and associates have examined the burrowing effects of wedge-tailed shearwaters (*Puffinus pacificus*) on several sandy cays in the Capricorn Group of islands near the southern end of Australia's Great Barrier Reef. These birds commonly nest at the end of a burrow that often exceeds 2 m in length (Fig. 4.2) (Dyer and Hill 1991). Burrows are aggregated (Dyer and Hill 1990) according to patterns of vegetation distribution and human disturbance; areas with dense *Pisonia* ground cover and branches are especially favored (Hill and Barnes 1989), and burrows are so aggregated that the resulting surface looks more like a bleak moonscape (e.g., Dyer and Hill 1990, fig. 2). Over thirty-five thousand breeding burrows were censused from only two small islands in the group, Heron and Masthead islands, and average burrow densities are ~0.12 m^{-2} (Hill and Barnes 1989). Burrow lengths on Heron and Erskine islands averaged around 0.91 m, but ranged from 0.10 to 2.35 m long, and were normally distributed (Dyer and Hill 1992). From diagrams in Dyer and Hill (1991), it appears that the average height and width dimensions of a wedge-tailed shearwater burrow are both approximately 10 cm. Assuming a 0.1 m × 0.1 m × 0.91 m burrow represents a normal-sized one, each burrow displaces 0.009 m^3 of sediment. For the islands of Heron and

Masthead, then, more than 318 m^3 of sediment have been excavated by burrowing wedge-tailed shearwaters.

Furness (1991) described the construction of burrows by Manx shearwaters (*Puffinus puffinus*) on the Rhum National Nature Reserve in west Scotland. Burrows there average about 7 cm in diameter and slightly over 1 m long, so that 4,000 cm^3 of soil and stones are excavated for each burrow, in turn encouraging further erosion by rain and wind. Furness believed, however, that this erosion was offset by vegetative stimulation induced by nitrogen and phosphorus deposition associated with the birds' guano.

Hall and Williams (1981) examined the burrows of petrels and prions at subantarctic Marion Island. The numbers of these birds there is unknown, but estimated at many hundreds of thousands to millions. Individual burrows there could cause removal of up to 1 m^3 of peaty soil, but Hall and Williams used a conservative estimate of 0.2 m^3 of material removed from each burrow. They multiplied this value times a range of birds of 0.6–1 million, and derived a figure for the quantity of material removed by burrowing birds on the order of $1.2-2 \times 10^5$ m^3, certainly one of the largest amounts of eroded materials attributed to birds anywhere on earth. Fugler et al. (1987) noted that feral cats on Marion Island are causing widespread devastation of the blue petrel (*Halobaena caerulea*) population, so that unless eradication efforts are undertaken, this significant geomorphic process is likely to be severely curtailed in the future.

Atlantic puffins (*Fratercula arctica*) often nest in large colonies, and most of these are on earthy islands where the birds can burrow. Over one-quarter million pairs breed along the coast of Newfoundland, and another eighty-seven thousand pairs occupy nest colonies on the island of Labrador (Snyder 1993). Burrow densities of Atlantic puffins in some subarctic locations are quite high: Harris and Birkhead (1985) reported instances of over three burrows per square meter. Densities are highest on sloping ground, because burrowing on level ground is restricted to one level with tunnels running parallel to the surface. In his study of Atlantic puffins at Great Island, Newfoundland, Nettleship (1972) found that burrow density was negatively correlated with distance from cliff edges, and positively correlated with angle of slope and soil depth. The three variables together accounted for over 81% of the variation in burrow abundance. On relatively flat islands, where slope variation is limited, soil depth is the main factor affecting puffin burrow density (Birkhead and Harris 1985).

Furness (1991) summarized the classic case of the basic destruction of an entire island off the coast of Wales by Atlantic puffins. In 1890 there were over half a million puffins on Grassholm, an 8.9-ha island; densities of breeding puffins were two or three pairs per square meter. Vegetation was nearly totally destroyed, and the soil was so severely tunneled by puffin burrows that it collapsed and was extensively eroded. By 1928, all that was left were isolated pillars and tussocks of turf, and about two hundred puffins.

Penguins are some of the largest and most numerous of the burrow-nesting colonial seabirds. Blackfooted penguins (*Spheniscus demersus*) number about a hundred and seventy-six thousand (Müller-Schwarze 1984), and burrow for protection from high subtropical insolation. LaCock (1988) showed the preference of blackfooted penguins for unsandy sites in choosing burrow sites (see also Müller-Schwarze 1984, fig. 71). Stokes and Boersma (1991) examined the effects of substrate on the distribution of Magellanic penguins (*S. magellanicus*) burrows in southern Argentina (see Müller-Schwarze 1984, fig. 76), and found that slopes and soils with intermediate mixes of sand, silt, and clay were most amenable to burrowing. A typical burrow for a Magellanic penguin has a relatively wide entrance that narrows to a short neck, then subsequently widens into a chamber where eggs are laid. Some tunnels may be >1 m in length, and as much as 1–2 m deep. Stokes and Boersma (1991) found that, in a sample of fifteen burrows, the mean length was 59.3 cm, width at entrance was 56.3 cm, width at neck was 37.3 cm, and height was 21.1 cm. An average burrow therefore required removal of ~0.05 m^3 of substrate. The rookery in question had over one million birds present, with between two hundred thousand and four hundred forty-six thousand breeding pairs (Müller-Schwarze 1984). If these estimates of breeding pairs are accurate, somewhere in the range of 10,000–22,000 m^3 of sediment have been excavated at this one rookery, and Müller-Schwarze (1984) notes that there are many more such breeding sites throughout the subantarctic region.

Conclusions

As a group, birds are probably much more geomorphically important than would be expected at first glance. The overall geomorphic effect of burrowing colonial seabirds is very significant, and may be the dominant geomorphic process of sediment removal at some puffin and penguin rookeries. Elsewhere, such as in the goose-affected salt marshes of northern Canada, birds are a major geomorphic influence on the local scene, and also exacerbate the effects of other processes, such as the work of waves and running water. Unfortunately, however, birds have been virtually ignored in the geomorphic literature. Only Hall and Williams (1981), Dionne (1985), Mitchell (1988), and Johnson (1993) stand out as geomorphologists, rather than biologists, whose work has addressed the geomorphic influences of birds. The transient nature of most midlatitude birds and the isolation of major seabird colonies from population centers have combined to lull most North American geomorphologists into believing in avian geomorphic insignificance – a conclusion that is simply not true!

5
The geomorphic effects of digging for and caching food

Mammals dig for a variety of reasons, but essentially these reasons echo those of other animals: to dig up food, either floral or faunal; to cache provisions; and to excavate habitations, whether temporary, seasonal, or permanent. This chapter examines the geomorphic effects of mammals as related to the obtainment and caching of food. Excavations associated with habitation are examined in Chapter 7.

Many mammals engage in geomorphically significant digging activities while searching for food. Once a food supply is acquired, a number of mammals subsequently produce geomorphically important effects by caching it. The following sections examine these dual roles of digging, in search and in storage of food.

Digging for food

Excavation of food by mammals occurs in pursuit of both plant and animal sources. Large herbivores such as ungulates paw at the surface to reveal succulent roots and forbs, whereas small herbivores may dig and/or tunnel extensively in pursuit of plant-food sources beneath the surface (Andersen 1987) (Fig. 5.1). Carnivores dig vigorously in pursuit of prey in burrows, and omnivores dig extensively for a wide variety of food. Mammals that produce geomorphically significant food excavations includes such diverse animals as aardvarks (Dean and Siegfried 1991), warthogs and other wild pigs (Tisdell 1982; Lacki and Lancia 1983; Sowls 1984), porcupines (Gutterman and Herr 1981; Yair and Rutin 1981; Gutterman 1982, 1987; Yeaton 1988; Alkon and Olsvig-Whittaker 1989; Gutterman, Golan, and Garsani 1990; Shachak, Brand, and Gutterman 1991), badgers (Knopf and Balph 1969; Messick and Hornocker 1981; Long and Killingley 1983; Neal 1986; Moore 1990), moles (Jonca 1972), pocket gophers (Teipner, Garton, and

Figure 5.1. Hoary marmot (*Marmota caligata*) digging for food in an alpine meadow; Glacier National Park, Montana.

Nelson 1983; Huntly and Inouye 1988; Thorne and Andersen 1990), baboons (Dean and Milton 1991a), bat-eared foxes (Dean and Milton 1991a), rabbits (Rutin 1992), bandicoots (Morcombe 1968), grizzly bears (Butler 1992), whales (Oliver et al. 1983b, 1984; Kvitek and Oliver 1986; Nelson and Johnson 1987; Nelson, Johnson, and Barber 1987; Weitkamp et al. 1992), walrus (Nelson and Johnson 1987; Nelson et al. 1987; Klaus, Oliver, and Kvitek 1990), and sea otters (Hines and Loughlin 1980; Kvitek and Oliver 1988; Kvitek et al. 1988).

Excavations may be spatially concentrated on slopes with a given steepness, aspect, or narrow elevational zone where a food source is concentrated. Examples include grizzly bears (*Ursus arctos horribilis*), which excavate large boulders on talus slopes of 31–40° above tree line, where aggregations of army cutworm moths occur, especially on west-facing slopes (Mattson et al. 1991); wild boars (*Sus scrofa*), which scour the soil at upper tree line with their snouts in search of voles (Martinez Rica and Pardo Ara 1990); and porcupines (*Hystrix indica*), whose digging actions may be ten to twenty times lower on south-facing slopes than on adjoining north-facing ones because of the paucity of soil cover on drier south-facing slopes (Yair and Rutin 1981).

There are also seasonal variations in the geomorphic effectiveness of mammals digging. In some cases, these variations result from periods of inactivity in the digging animal, as in the case of winter hibernation by grizzly bears (Craighead and Craighead 1972). In other instances, it is the prey food that is seasonally inactive, a prime example being termites: These are less active in their termitaria during the cooler winter season, and congregate at greater depths, such that they are less accessible to the digging efforts of aardvarks (Willis, Skinner, and Robertson 1992). The aardvarks respond by pursuing alternative food sources, thereby reducing their geomorphic effectiveness. Willis et al. (1992) also summarized literature that illustrated a similar lower utilization of termites by aardvarks during dry seasons, when the termitaria presumably prove more resistant to excavation.

It would be impractical to describe in detail the digging habits and geomorphic results of every mammal in pursuit of food; instead, I present here a series of examples from the literature, in which widespread, geomorphically significant and quantifiable excavations for food are created. At the same time, I urge readers to remain cognizant of the other widespread mammalian food-excavating activities for which few or no quantitative data exist.

Spiny anteaters

Mitchell (1988) described the geomorphic effects of the echidnas, or spiny anteaters (*Tachyglossus aculeatus*), of Australia. The echidna is found in all Australian life zones, from subalpine regions to the tropics; Mitchell's detailed work on the geomorphic effects of their feeding was carried out in New South Wales, ~245 km WNW of Sydney. Spiny anteaters excavate vertical faces into surface soil or directly into ant mounds. Single scrapes (all data from Mitchell 1988) are 15–40 cm long, 12–25 cm wide, and 2–20 cm in depth, with a mean size of 22 × 16 cm in area and 6 cm deep (i.e., 2,112 cm^3 in volume). From twelve study plots, Mitchell calculated a soil-disturbance rate of 1.65 t ha^{-1} yr^{-1}. He also noted that bandicoots (*Isodon* spp.), wombats (*Vombatus ursinus*), and gray kangaroos (*Macropus major*) produced significant but unquantified amounts of soil scraping and bioturbation. The overall net *erosional,* not bioturbational, role of these animals and the spiny anteater was not examined.

Porcupines

The bulk of the research carried out on the geomorphic effects of porcupines has been done by Israeli scientists studying excavation sites created in the Negev Desert highlands by the Indian crested porcupine (*Hystrix indica*). There, porcupines excavate and consume corms, bulbs, tubers, and underground parts of leaves and stems (Gutterman 1982, 1987). Porcupines may dig as deeply as 25–30 cm in search of such food (Gutterman 1987; Gutterman et al. 1990), but depths of 10 cm seems most common. The density of diggings varies greatly with variations in soil moisture and plant-food availability. Gutterman and Herr (1981) reported excavation densities of up to one digging every 2 m^2 per year, and Alkon and Olsvig-Whittaker (1989) reported a range of over one dig per square meter on a densely vegetated site to only 0.1 per square meter on sparsely vegetated sites. Gutterman et al. (1990) cited the highest density of three new diggings per square meter.

Excavations by porcupines tend to be elongated depressions, with soil ejected by the porcupines' toes in a downslope direction. The volume of individual excavations ranges from ~400 cm^3 to as high as 1,900 cm^3 (Gutterman and Herr 1981; Yair and Rutin 1981; Alkon and Olsvig-Whittaker 1989; Shachak et al. 1991), although some new diggings as small as 100–170 cm^3 have been reported (Alkon and Olsvig-Whittaker 1989). Shachak et al. (1991) counted a total of 6,609 new diggings in a fourteen-year time span in their study area, with an average size of 257 cm³. Using these data, I calculate that 1,698,513 cm³ of sediment were excavated by porcupines during that period, or an average of 121,322 cm³ yr⁻¹.

In terms of the amount of total sediment disruption per surface area, Yair and Rutin (1981) estimated mean soil displacement by porcupines of 2.9–8.7 g m⁻² yr⁻¹. Alkon and Olsvig-Whittaker (1989) found slightly higher annual soil-displacement rates, ranging from 3.6 to 18.2 g m⁻². The total area disturbed in the Alkon and Olsvig-Whittaker study ranged from 0.7 to 3.6% on 1,000-m² plots; from these figures they calculated that up to 0.8% of the total surface area may be disturbed by porcupines annually.

Because the porcupine excavations are concentrated on hillslopes, they eventually in-fill with sediment washed in from above and blown in. The depressions also serve as moisture-collection sites that aid in the reestablishment of plants, which in turn entice subsequent porcupine excavations, suggesting the existence of a mutualism between the porcupines and the plants dependent on favorable growth sites, that is, porcupine excavations (Gutterman 1987). A minor controversy currently exists as to the recovery rate of individual porcupine diggings. Gutterman and associates site several examples where diggings may take up to twenty years to recover (Gutterman 1987; Gutterman et al. 1990; Shachak et al. 1991), whereas Alkon and Olsvig-Whittaker (1989) provided much shorter lifespan estimates, in the range of 1–6.5 years. The differences may be a function of the methodologies employed: Shachak et al. (1991) based their estimates on long-term field observations (14 yr); Alkon and Olsvig-Whittaker's (1989) estimates extrapolated via regression analysis the results of a seventeen-month sampling period. Whatever the truth concerning recovery rates may be, it is indisputable that porcupines are one of the singularly most significant geomorphic agents in the Negev Desert highlands, where "almost all the soil surface can be covered by porcupine diggings in different stages of being filled in" (Gutterman et al. 1990, p. 122).

Badgers

Although a great deal has been published on badger burrow complexes, or "setts" (see Chapter 7), few quantitative data have been forthcoming concerning the amount of sediment displaced in pursuit of food. Platt (1975) provided a description of the localized disturbances caused by the American badger (*Taxidea taxus*) while digging for ground squirrels in their burrows at night. Platt's research was focused on patch disturbance dynamics, rather than the geomorphological effects of badger excavations, but he noted that badgers produced a hole and corresponding mound of soil ~0.2–0.3 m² in size in pursuit of ground squirrels. He also noted that the mounds slowly disappeared over about twenty years, as slow succession to undisturbed prairie occurred.

Wild pigs

In Australia, wild (feral) pigs wreak extensive economic havoc on the delicate semiarid and arid landscape by their widespread rooting for food. Although he did not provide quantitative data, Tisdell (1982) vividly demonstrated their geomorphic effectiveness. Pig rooting undermines earthen dams and fences, and public dirt roads have been rendered useless by the action. A light-aircraft landing strip in Queensland was "entirely rooted up by pigs and rendered completely unserviceable for aircraft" (Tisdell 1982, p. 34). The owner of one property of 200 square miles in Queensland reported that fully 10% of his land (20 square miles) was rooted by pigs. Similar levels of destruction have been reported from the humid forested environment of the Luxembourg Ardennes; there, Imeson (1976, p. 115) reported that rooting by wild pigs displaced "upwards of a ton of earth" over the course of a few days, leaving a characteristic microtopography of mounds and pits ~40–70 cm deep (Imeson 1977).

Aardvarks

The aardvark (*Orycteropus afer*) is a nocturnal burrowing mammal, native to Africa, that feeds primarily on ants and termites. The latter especially are consumed via excavation of termitaria; ants are consumed both on the surface and by ant-mound excavations (Willis et al. 1992).

Dean and Siegfried (1991) examined 344 aardvark diggings into termite mounds in the eastern Karoo, South Africa. They found that the diggings were spatially concentrated on west-facing portions of the termitaria. Their results contrast with those of Bernard and Peinke (1993), who concluded that aardvarks concentrate their diggings on the northern and eastern quadrants of termite mounds.

Aardvark diggings create patch environments for germination by plants by loosening the soil and increasing the soil's water-retention capabilities (Dean and Milton 1991b). Dean and Milton (1991a) described the morphology of diggings created by aardvarks as shallow, basin-shaped scoops with a generally oval shape. They reported a concentration of over ninety diggings per hectare. Morphometrically, the diggings were on average 16 cm long, 8 cm in width, and ranged from 5 to 7 cm in depth. Each therefore eroded ~640–900 cm^3 of sediment (volume values calculated from the data in Dean and Milton 1991a); so with more than ninety diggings per hectare, a minimum of 57,600–81,000 cm^3 ha^{-1} of sediment is excavated by aardvarks in optimal locations with termitaria.

Grizzly bears

The grizzly bear (*Ursus arctos horribilis*) is a large omnivore, weighing 160 kg or more and standing 1.25 m at the shoulder (Butler 1992). It was formerly distributed across the western half of North America (Interagency Grizzly Bear Committee 1987), but its current distribution is largely restricted to the western Canadian cordillera and Alaska. It is also found in isolated locations in circumpolar regions of Eurasia (subspecies *Ursus arctos arctos*), but few descriptions of geomorphic effects of these bears have been forthcoming (see Ustinov [1976] for an exception). The grizzly bear is a threatened species under the U.S. Endangered Species Act in the lower forty-eight United States, where fewer than a thousand bears, occupying <1% of their pre-European range (Wilkinson 1993), still exist. The bulk of these bears are located in two regions: the Yellowstone ecosystem of northwestern Wyoming and surrounding areas, where perhaps 350 bears live (Interagency Grizzly Bear Committee 1987), and the Northern Continental Divide ecosystem of northwestern Montana, with perhaps nearly six hundred bears in residence. Approximately two hundred of these bears live in Glacier National Park (Martinka 1971, 1972, 1974a–c; Butterfield and Key 1986; Keating 1986, 1989; Martinka and Kendall 1986; Hayward 1989), with the remainder distributed throughout adjacent federal wilderness areas and state and local lands.

The most distinguishing characteristic of a grizzly bear from a geomorphic perspective is the prominent hump of massive muscles over the shoulder (Fig. 5.2). This muscular mass, in concert with impressive front claws reaching 10 cm in length (see Fig. 6.1), makes the bear a truly formidable digging machine. The omnivorous grizzly eats a wide variety of food types, including animal carrion, ungulates (deer, elk, and moose), rodents, fish, insects (especially ants), pine nuts, berries (especially huckleberries), green vegetation, and a variety of bulbs and tubers (Singer 1976, 1978b; Kendall 1983; Martin 1983; Holcroft and Herrero 1984; Butterfield and Key 1986; Herrero, McCrory, and Pelchat 1986; Mace and Bissell 1986; Mace and Jonkel 1986; Martinka and Kendall 1986; Schoen, Lentfer, and Beier 1986;

Figure 5.2. The North American grizzly, or brown, bear. The large, humped shoulder provides massive strength for digging in concert with long, sharp claws. Photo courtesy of Eugene Palka.

Hamer and Herrero 1987; Edge, Marcum, and Olson-Edge 1990; Hamer, Herrero, and Brady 1991; Mattson, Blanchard, and Knight 1991, 1992; Mattson et al. 1991). Geomorphically significant excavations are produced by the bears especially in search of rodents, bulbs and tubers, and aggregations of army cutworm moths.

Although not all items consumed by grizzly bears require digging, grizzlies do dig throughout their period of activity from April to November. During the spring, much activity is concentrated along riparian corridors (Singer 1978b; Malanson 1993). During summer months, great amounts of digging take place above upper timberline and in the alpine tree-line ecotone (Butler 1992).

In their search for succulent roots and tubers, as well as pine nuts, grizzlies excavate broad areas to shallow depth. Those that occur in riparian habitats may not have lasting geomorphic significance; there, the absence of a steep slope results in considerable backfilling of excavations (Butler 1992). The net effect is then one of pedoturbation, *sensu* Johnson (1989, 1990, 1993; Johnson et al. 1987; Johnson and Balek 1991), not denudation.

Diggings for plant bulbs and tubers on steeper slopes such as shown in Figure 5.3 are widespread in the upper tree-line ecotone during summer. In this environment, glacier lily (*Erythronium grandiflorum*) bulbs (Butler 1992; Fig. 5.4) and yellow sweetvetch (*Hedysarum sulphurescens*) roots are major food sources. I have seen a hillslope on the eastern flank of East Flattop Mountain in Glacier National Park, Montana, where fresh excavations covered an area at least 75 m long and 25 m across. Such excavations are enormously impressive to encounter in the field, and certainly add velocity to one's own movements!

Yellow sweetvetch is widespread in the southern Canadian and Montana Rockies, and grizzly bears extensively excavate for the roots (Holcroft and Herrero 1984; Herrero et al. 1986; Edge et al. 1990; Hamer et al. 1991). Ease of digging may be the most important factor in determining where and how extensively grizzly bears will excavate yellow sweetvetch (Holcroft and Herrero 1984; Edge et al. 1990), more important than the abundance of the plant. Excavations were particularly noted on steep slopes underlain by gravelly and rocky substrate (Holcroft and Herrero 1984); such site conditions are amenable to sweeping of the overburden downslope by the bear's powerful paws in order to reveal the buried roots.

My own work (Butler 1992) provided one of the few quantitative measures of the amount of sediment excavated by grizzly bears in association with the search for glacier lily bulbs. I measured several such excavations

Figure 5.3. Typical alpine tree-line habitat frequented by grizzly bears. Digging for flower bulbs and corms, as well as for ground squirrels, is common in this ecotone; Highline Trail, Glacier National Park, Montana.

(Fig. 5.4) in the field in Glacier National Park, Montana, and assumed that the two hundred grizzlies there each produce five such excavations per year. (See Butler [1992] for details and caveats of the methodology.) Each excavation is ~0.3 m^3 in volume, and so 200 bears yield 1,000 excavations × 0.3 m^3 = 300 m^3 yr^{-1} of geomorphic displacement downslope. Thus, over the past hundred years, during most of which time the grizzly population of Glacier National Park has been protected and relatively stable (Martinka 1971; Hayward 1989), a minimum of 30,000 m^3 of sediment has been displaced into the debris cascade by grizzly bears digging for plant bulbs on steep slopes in the alpine tree-line ecotone. Given the widespread additional excavations in Glacier National Park and elsewhere for additional buried plant-food sources – in Yellowstone National Park, entire hillsides may be ripped up by bears in search of the western spring beauty, *Claytonia lanceolata* (McNamee 1990) – it can be seen that this value indeed represents only a barely minimal representation of the geomorphic work produced by grizzly bears.

In addition to this large amount of geomorphic work carried out in pursuit of plant-food sources, consider also the additional excavations created by grizzly bears digging for rodents such as marmots and ground squirrels (Bailey and Bailey 1918; Mills 1919; Gardner, Smith, and Desloges 1983; Butler 1992; Slaymaker 1993) (Fig. 5.5), and for insect aggregations of ladybird beetles and army cutworm moths (Chapman, Romer, and Stark 1955; Mattson et al. 1991; Gilmore 1993). Excavations for army cutworm moths and other insects are dug during summer months at alpine sites in Montana and Wyoming (Mattson et al. 1991; Gilmore 1993), on active scree slopes above the tree line. Rocks and boulders are literally shoved and tossed downslope in pursuit of the insect bonanzas, a fact that should not go unconsidered by geomorphologists studying alpine talus-slope processes and clast movement (*sensu* Gardner et al. 1983) in the North American cordillera. Typical excavations for insects are 1.5–5 dm deep, as reported by Mattson et al. (1991), who could not consistently determine total volumes excavated because bears often backfilled previous excavations as they created new ones.

Similar extensive, frenzied digging has been reported for grizzlies in pursuit of rodents. In 1919, Mills wrote:

The grizzly eagerly earns his own living; he is not a loafer. Much work is done in digging out a cony, a woodchuck, or some other small animal from a rock-slide. In two hours' time I have known him to move a mass of earth that must have weighed tons, leaving an excavation large enough for a private cellar. I have come upon numbers of holes from which a grizzly had removed literally tons of stone. In

Figure 5.4. Typical grizzly bear excavation for underground bulbs of the glacier lily (*Erythronium grandiflorum*); Glacier National Park, Montana.

(a)

(b)

Figure 5.5. (a) Grizzly bear excavation of a Columbian ground squirrel den site, on a 30°–35° slope; Glacier National Park, Montana. (b) Grizzlies digging for ground squirrels in Denali National Park, Alaska. Photo (b) courtesy of Eugene Palka.

places these holes were five or six feet deep. Around the edges the stones were piled as though for a barricade. In some of them several soldiers could have found room and excellent shelter for ordinary defense. (p. 71)

Later in the same work, Mills (p. 125) again referred to the extraordinary excavational capabilities of the grizzly bear: ". . . he stopped and dug energetically. Buckets of earth flew behind, and occasionally a huge stone was torn out and hurled with one paw to the right or left."

Excavations for rodents are typically on slopes sufficiently steep so that the excavated material is deposited up to several meters downslope. In order to quantify the effects of grizzly bears digging for ground squirrels and other rodents, I again used Glacier National Park, Montana, as an example (Butler 1992), because of its firm grizzly bear population estimates (see above). Published descriptions and field observations (see Fig. 5.5) reveal the volume of the typical excavations for burrowing animals to be perhaps 0.2 m^3 per site (Butler 1992). I assumed that the two hundred grizzlies of Glacier National Park each produce five excavations per year in search of rodents, yielding a probably minimal estimate of 200 m^3 yr^{-1} excavated by the bears, or approximately 20,000 m^3 over the past hundred years (Butler 1992). Combined with the excavations for plant food and the unquantified effects of "boulder tossing" in search of insects on alpine talus slopes described above, grizzly bears in Glacier National Park alone have excavated a minimum of 50,000 m^3 of slope materials in the past century, the bulk of these materials being eroded above or within the alpine tree-line ecotone. If, as some studies suggest (Interagency Grizzly Bear Committee 1987), the pre-European, lower-forty-eight-state grizzly bear population was in excess of a hundred thousand, it then takes little imagination to envision the truly vast geomorphic contributions the bears formerly made across western North America. Fortunately, they still do so in favored places even as the twentieth century draws to a close.

The California gray whale

One of the most well-documented, and certainly most unusual, examples of the geomorphic significance of mammalian feeding is associated with the California gray whale (*Eschrichtius robustus*). Gray whales calve and breed in lagoons of Baja California in winter, and migrate to feeding grounds in the Bering and Chukchi seas in summer (Rice and Wolman 1971; Nerini and Oliver 1983; Johnson and Nelson 1984; Klaus et al. 1990). Feeding has also been reported in high concentrations along the west coast of Vancouver Island, British Columbia (Murison et al. 1984; Oliver et al. 1984; Kvitek and Oliver 1986), in modest numbers in the Klamath River estuary of northern California (Avery and Hawkinson 1992), and along the beach of Puget Sound, north of Seattle, Washington (Ezzell 1991; Weitkamp et al. 1992).

The gray whale is unique among large cetaceans in that it feeds almost exclusively upon benthic (bottom-dwelling) organisms (Nerini 1984), especially amphipod crustaceans (Oliver et al. 1983b, 1984). The whales ingest sediment and fauna from the ocean floor, apparently through the action of suctioning or "slurping" deep furrows into the ocean floor (Nerini 1984; Kvitek and Oliver 1986; Obst and Hunt 1990). The sediment is filtered through the whales' baleens to remove the invertebrates, and subsequently expelled near the surface as large mud plumes (Harrison 1979; Oliver et al. 1983b, 1984; Obst and Hunt 1990). Although this seafloor excavation typically occurs at depths of several to tens of meters beneath the surface, Ezzell (1991) and Weitkamp et al. (1992) described gray whale feeding pits created at depths of less than 4 m along the Puget Sound coastline, such that the excavations were actually visible in the littoral zone at low tide.

In the process of excavating and expelling sediment in order to extract benthic fauna, gray whales also actually ingest large amounts of sediment, which is subsequently transported in their stomachs. Rice and Wolman (1971) report several cases where individual whales transported 10–100 kg of sediment in this manner.

Gray whale feeding excavations vary in size according to the size of the whale performing the excavation. They are primarily carried out in fine-grained silts and clays, the benthic substrate most amenable to the primary food of the whales. The excavations tend to be oblong in shape, with volume removed ranging from ~0.37 m^3 (Johnson and Nelson 1984; Nerini 1984; Cacchione et al. 1987; Nelson and Johnson 1987; Nelson et al. 1987) to 1.12 m^3 (value calculated from data in Avery and Hawkinson [1992]). A single whale may make as many as six such excavations during one dive (Oliver et al. 1984), and double excavations resulting in striking heart-shaped pits are apparently not uncommon (Fig. 5.6) (Oliver et al. 1984; Avery and Hawkinson 1992). Pits made in one feeding season ("fresh" pits) are distinguishable from enlarged pits, scoured by seasonal storms, from previous feeding seasons (Nelson and Johnson 1987; Nelson et al. 1987).

Weitkamp et al. (1992) described an estimated 2,700–3,200 gray whale feeding pits from a 90-km coastal sand flat along Puget Sound during the 1990 field season, and nineteen thousand pits from a 180-km coastal stretch during their 1991 field season. The volume of the average pit varied by year: Pits excavated in 1990 averaged ~0.37 m^3, whereas the larger 1991

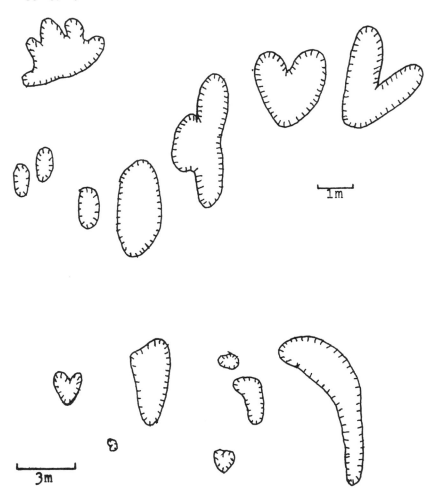

Figure 5.6. Plan views of typically shaped gray whale excavations, including heart-shaped double excavations. Redrawn from Oliver et al. (1984) and Avery and Hawkinson (1992).

pits averaged 0.65 m³ (values calculated from data presented in Weitkamp et al. [1992]). Over the 90-km coastal study area of 1990, 999–1,184 m³ of sediment were excavated, or 11–13 m³ km⁻¹. In 1991, the larger pits and greater numbers provided a total of 12,350 m³ excavated along 180 km of coastline, or 68–69 m³ km⁻¹. Clearly, gray whales can produce significant impacts over a localized area such as along the Puget Sound coastline. Other studies described below reveal that their influence is even more pervasive and widespread.

Kvitek and Oliver (1986) used side-scanning sonar to determine the extent of ocean floor disturbed by gray whale feeding in Ahous Bay, Pachena Bay, and Port San Juan, Canada. A range of roughly 18–22% of the ocean floor in these areas was disturbed by gray whale feeding. Unfortunately, no data on the actual amount of sediment disturbed or removed were provided.

Nelson and Johnson (1987) and Nelson et al. (1987) attributed astonishingly high amounts of geomorphic work to gray whales feeding on the shelf of the Bering Sea. The approximately sixteen thousand whales there covered about 5.6% of the Bering shelf feeding area with new feeding excavations in one year, an area of ~1,200 km^2. By assuming an average pit depth of 10 cm in concert with their pit area measurements from bottom samples, bottom photographs, underwater video, side-scan sonar, and scuba-diver observations (Nelson et al. 1987), they calculated that whale feeding annually resuspends a minimum of 120 million m^3 of sediment there! This value is nearly three times the annual load of suspended sediment discharged into the Bering Sea by the Yukon River, the fourth largest sediment source in North America (Johnson et al. 1987; Nelson and Johnson 1987). They also stressed that this resuspended sediment does not simply settle back to the seafloor, but is carried by northerly currents toward the Chukchi Sea.

The Pacific walrus

In the northern hemisphere, the Bering Sea supports by far the largest variety and abundance of pinnipeds; however, only two of these pinniped species, the bearded seal (*Erignathus barbatus*) and the Pacific walrus (*Odobenus rosmarus*), feed primarily on benthic invertebrates (Lowry, Frost, and Burns 1980). Bearded seals feed primarily on clams, crabs, shrimp, and cockles, but little is known concerning the quantitative geomorphic effects of bearded seal predation on benthic fauna. Walrus, however, have been studied more extensively and quantitatively.

Walrus feed by swimming along the ocean floor with their heads down, looking for prey or sensing it with "vibrissae," whiskerlike projections that cover the snout (Nelson and Johnson 1987; Nelson et al. 1987). "Rooting" with the snout is used to excavate shallowly buried prey, and more deeply buried prey is excavated by hydraulically jetting pulses of water in a fashion similar to that described for some rays (Gregory et al. 1979). Although they have the ability to dive to depths of >100 m, they have little reason to do so, because most of their food sources are on shallow ocean shelves at depths of <30 m (Wiig et al. 1993).

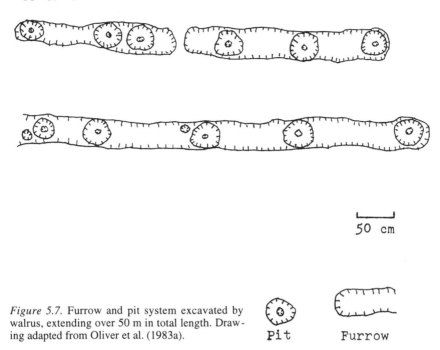

Figure 5.7. Furrow and pit system excavated by walrus, extending over 50 m in total length. Drawing adapted from Oliver et al. (1983a).

Through its snout rooting and water jetting, a walrus creates two distinct forms of excavation (Fig. 5.7). Snout rooting creates long furrows of a width approximately equal to that of typical walrus snouts (Oliver et al. 1983a), that is, about 45 cm; furrow depths average ~17 cm, and furrows of 4–40 m or longer have been reported (Oliver et al. 1983a; Nelson et al. 1987; Klaus et al. 1990). The hydraulic jetting of water creates roughly circular excavations, or pits, typically 30 cm across and 30–35 cm deep (Oliver et al. 1983a; Klaus et al. 1990).

Like the gray whale, walrus can disturb geographically widespread parts of the ocean floor; estimates of up to 40% were summarized by Klaus et al. (1990). Oliver et al. (1983a) described walrus feeding grounds in the Bering and Chukchi seas, on the fringes of gray whale feeding areas and overlapping with those of bearded seals. An estimated walrus population of two hundred thousand inhabits the region. Oliver et al. (1983a) estimated that 6–11% of the ocean floor around Cape Nome, Alaska, was disturbed by walrus feeding, and also provided the 40% estimate cited by Klaus et al. (1990) for the area surrounding Sledge Island.

Nelson and Johnson (1987) and Nelson et al. (1987) provided the most detailed quantitative estimate of walrus' influence on the geomorphology of the Bering shelf. They assumed that the area's two hundred thousand walrus spend approximately a hundred days in the area annually, and that each walrus digs a minimum of two furrows a day of (their) average dimensions of 47 m long × 0.4 m wide × 0.1 m deep. Their estimates therefore did not even include the additional effects of the widespread pit excavations of ≥30 cm depth. Nevertheless, their calculations revealed that walrus disturb about seventy-five million cubic meters of sediment annually. Unlike the case of the gray whale, however, these disturbances occur in coarser, sand-rich sediments that do not travel great distances upon disturbance, but typically settle back to the seafloor near the point of origin. Nelson et al. (1987) also did not observe widespread current-scour enlargement of the walrus furrows, such as typifies the gray whale excavation pits. Even so, this is a tremendous amount of annually produced bioturbation that results in widespread ocean-floor modification and creation of patch habitats (Oliver et al. 1983a).

The sea otter

The sea otter (*Enhydra lutris*) forages in the benthos of both rocky and soft-sediment communities (Riedman and Estes 1988), but its geomorphic effects are primarily restricted to the latter. It was nearly wiped out by fur trading in the 1800s, but has made a strong recovery in numbers in the latter half of the twentieth century. Its current distribution ranges from the Kuril Islands and Kamchatka Peninsula through the Aleutian Islands up to Prince William Sound, Alaska, and off the Monterey coast of central California (Riedman and Estes 1988). The sea otter feeds primarily on benthic bivalves, especially favoring the shallow-burrowing pismo clam (*Tivela stultorum*) (Calkins 1978; Hines and Loughlin 1980), some deep-burrowing clams (*Tresus nuttallii* and *Saxidomus nuttalli*) (Kvitek and Oliver 1988; Kvitek et al. 1988), and mussels (family Mytilidae) (VanBlaricom 1988). Sea urchins associated with offshore kelp communities are also favored by the sea otter (Foster and Schiel 1988).

Sea otters affect the geomorphology of offshore communities in two ways, both associated with feeding. The first falls primarily into the minor curiosity realm. The sea otter picks up and transports stones from the seafloor to use as tools in opening shells (Hall and Schaller 1964; VanBlaricom 1988). The sea otter thus joins the chimpanzee (which uses stones to crack open nuts; Boesch and Boesch 1981, 1984) as one of the few mammals to transport sediment for use as a tool to gain access to food sources. After picking up a stone, the otter places it across its chest while swimming on its back, thus using the stone as an anvil on which to crack open shells. Size, rather than particle shape, seems to be the criterion on which otters choose their cracking stones; Hall and Schaller (1964) reported that most stones averaged ~12 cm in diameter, and weighed in the range 468–666 g, although stones of up to 3.5 kg were excavated, transported, and subsequently abandoned as too large. Tool use is apparently more common among California sea otters, where a lesser abundance of prey requires greater effort in securing food than is necessary farther north (Riedman and Estes 1988).

The second way in which sea otters affect offshore geomorphology is through extensive excavations for clams and other burrowing shellfood. Hines and Loughlin (1980) examined the bottom topography in Monterey Harbor, California, where sea otters extensively excavate clams. They reported that bottom topography "was hummocky in these areas, and there were many craters 0.5–1.0 m across and 10–15 cm deep" (p. 160). By diving, they observed otters digging pits with their forelimbs like dogs; some pits were subsequently enlarged into trenches by rolling repeatedly from side to side. One such trench was >1.5 m long, 0.5 m deep, and 0.5 m wide, that is, ~0.4 m^3 in volume. Calkins (1978) reported clam-excavation pits created by sea otters in the subtidal and intertidal flats of Prince William Sound; there, pits were only about half the size of those reported by Hines and Loughlin (1980).

By far the most detailed quantitative data concerning sea otter excavations come from the work of Kvitek and associates (Kvitek and Oliver 1988; Kvitek et al. 1988). They described otter excavations in the Elkhorn Slough area of Monterey Bay, California. Pits were dug for deep-burrowing clams and for the shallow-burrowing Pismo clam. Pits on the floor of the slough were ubiquitous, with mean pit surface area of 1.4 m^2, and depths ranging typically from 30 to 50 cm, for pits dug to extract deep-burrowing clams. Using their data, I calculate that the average volume of a pit excavated for deep-burrowing clams is 0.4–0.7 m^3. It is not, unfortunately, known how many such sea otter pits exist in this or other areas, and how the number of pits excavated may vary from year to year. Kvitek et al. (1988) also described the shallow excavation pits created by sea otters in search of Pismo clams. They estimated that about twenty-five thousand Pismo clams were eaten over a 2.5-month period, and that sediment excavated per Pismo clam amounted to 0.01 m^3. Although each individual pit was small, multiplication by twenty-five thousand reveals that sea otters excavated 250 m^3 of sediment in only 2.5 months, in one small study area in Monterey harbor. When extrapolated over long times and greater geographic areas, such data provide a stronger picture of the geomorphic importance of sea otter excavations.

Caching of food

Assume now that a mammal has secured more food than it can consume, or that it must eat under conditions of drought, snow, or extremely cold temperature when food is not available. How does the animal deal at once with present food glut and upcoming unavailability? The answer, of course, is by storing (i.e., caching) the food for future use.

Anyone who has ever watched the family dog bury a bone has observed geomorphically effective food-caching behavior. The caching of food occurs throughout most mammalian families (Table 5.1), but not all caching is geomorphically significant. Large cats, for example, cache food in trees (Vander Wall 1990). Only those mammals known to cache food by digging burrows or by covering with soil raked from the surrounding area are listed in Table 5.1. Readers interested in the full range of food-caching site types and activities are referred to the excellent recent volume by Vander Wall (1990).

Many rodents cache food in extensive underground burrow systems (Vander Wall 1990). Attempting to distinguish the geomorphic effects of digging to cache food in a burrow system from digging a burrow for denning and protection is, however, virtually impossible. I therefore collectively address the geomorphic impacts of burrow and den construction in Chapter 7.

Five families of terrestrial carnivores cache food, but of those, only the Canidae (foxes, wolves, dogs), Felidae (tigers, bobcats, leopards), Ursidae (bears), and Mustelidae (weasels, mink, wolverines) geomorphically alter the surface. Canids typically excavate and bury their prey in shallow surface pits (Vander Wall 1990), but I know of no quantitative measurements of the geomorphic or pedoturbational effects of these actions. The Mustelidae cache prey in dens and burrows (see Chapter 7), and in some cases bury it in shallow surface pits similar to those of canids (Vander Wall 1990). Again, no quantitative measures of the effects of these actions are known. Bears and cats do not dig holes in which to bury prey, but instead rake soil and plant litter over it. The quantity of debris used by bears to cover prey is typically greater than used by cats, probably a result of the typically larger sized prey of bears (Vander Wall 1990).

Grizzly bears in many parts of their range, including central Europe, the Pyrenees, Russia, Finland, Norway, and North America, are known to scrape and rake large quantities of material over their prey (Elgmork 1982). McNamee (1990) provided vivid but unquantified descriptions of the raking process. Grizzly bears in Norway have been reported to scrape large areas

Table 5.1. *Mammals known to cache food in underground burrows
or by covering with soil*

Order	Family	Common name	Burial method
Insectivora	Soricidae	Shrews	Burrow
	Talpidae	Moles	Burrow
Carnivora	Canidae	Foxes, wolves, dogs	Burial in soil
	Ursidae	Bears	Soil raked over food
	Mustelidae	Weasels, mink, et al.	Burrows and burial in soil
	Felidae	Tigers, bobcats, et al.	Soil raked over food
Lagomorpha	Ochotonidae	Pikas	Burrow
Rodentia	Aplodontidae	Mountain beavers	Burrow
	Sciuridae	Squirrels & chipmunks	Burrow, burial in soil
	Geomyidae	Pocket gophers	Burrow
	Heteromyidae	Kangaroo rats, et al.	Burrow, burial in soil
	Cricetidae	Mice, hamsters, et al.	Burrow
	Spalacidae	Mole-rats	Burrow
	Rhizomyidae	Mole-rats	Burrow
	Arvicolidae	Voles & muskrats	Burrow
	Muridae	Old World rats & mice	Burrow, burial in soil
	Dipodidae	Jerboas	Burrow
	Hystricidae	Old World porcupines	Burial in soil
	Dasyproctidae	Agoutis & acouchis	Burial in soil
	Octodontidae	Octodonts	Burrow
	Ctenomyidae	Tuco-tucos	Burrow
	Bathyergidae	African mole-rats	Burrow

Source: Data from Vander Wall (1990).

around carcasses, averaging 43 m^2 (Elgmork 1982). Elgmork (1982) described the quantity of debris covering four carcasses as ranging from 0.5 to
1.5 m^3, with a suggestion that these values may be underestimates. Even so,
they are virtually the only known published quantitative data for the geomorphic effects of food caching by nonburrowing mammals. This is clearly
an area where additional research is needed, particularly by geomorphologists working with landscape ecologists attempting to understand patch dynamics. The accomplishments of the Israeli scientists in their studies of the
dynamics of porcupine excavations and plant patches would serve as an
example for others to follow, both in studies of animal excavations and of
food-cache sites.

6
Trampling, wallowing, and geophagy by mammals

As we have seen so far, the geomorphological effects of animals are both indirect and direct. The indirect effects of mammals, as well as other animals, are biotic, through the destruction of vegetation cover and litter, in turn altering community structure and possibly reducing ground cover, leading to enhanced erosion rates (McNaughton, Ruess, and Seagle 1988; Naiman 1988; Thomas 1988). Direct, abiotic effects (terminology from Thomas 1988) include those associated with digging, wallowing, trampling, and drinking. I have already described the extensive geomorphological results of mammals excavating for, and caching, food. The geomorphological results of digging for denning and habitation construction are examined in Chapter 7. In the present chapter, I focus on the geomorphic effects of trampling, wallowing, and geophagy.

It is difficult to separate the geomorphological effects associated with these processes. In order to drink, mammals may trample the landscape around a waterhole. In the process, the waterhole-fringe soils become compacted, leading to subsequent surface ponding of rainfall and enhancement of the waterhole, in turn attracting more animals. These animals in turn may wallow in the mud around the fringes of the waterhole, and/or consume parts of saline-laden soils surrounding waterholes that seasonally dry up. As the following sections are encountered, readers are urged to remember the complex interactions (*sensu* McNaughton et al. [1988] and Whicker and Detling [1988a] for indirect actions, and Thomas [1988] for direct) of these individual processes.

Trampling

Trampling by animals can lead directly or indirectly to erosion. It can be a direct agent of erosion, such as when trampling along the edge of a stream, pond, turf terrace, or erosion pan causes hoof or paw chiseling (Fig. 6.1)

Figure 6.1. Soil trampled by grizzly bear paw. Lens-cap diameter is 49 mm.

and bank sloughing and erosion (Buckhouse, Skovlin, and Knight 1981; Perez 1992b, 1993; Kondolf 1993). Trampling can also be an indirect agent of erosion because it "prepares" the soil for erosion by subsequent geomorphic processes. The indirect effects of trampling are difficult to quantify but can be categorized as the reduction or complete removal of the protecting vegetation cover, and increase in the bulk density of the soil (Fig. 6.2). The effects of trampling are frequently visually dramatic, as distinct ecological "edges" and "patches" are created by its effects (Dean and Milton 1991b; Reichman, Benedix, and Seastedt 1993).

Direct effects of trampling are virtually unquantified in the geomorphic literature. Dean and Milton (1991b) noted that antelopes sheltering under trees during the heat of the day trample soil and also enrich it with fecal pellets. Natural animal trails produced by trampling are widespread and are frequently the subject of nonquantitative comment in the literature (Fig. 6.3a,b). The animals that have produced them and the environments in which they are created include hippopotamuses and elephants in tropical grasslands (Lock 1972, Haynes 1991), bison in the semiarid grasslands of the Great Plains (Fig. 6.4) (Clayton [1975, 1976], although Babcock [1976] disputed Clayton's evidence), mountain goats in the Rocky Mountains (Eaton 1917; Butler 1993), and caribou in the arctic tundra of the Barren Grounds of Canada's Northwest Territories (Harper 1955). Clayton (1975, 1976) noted the widespread nature and number of American bison trails in the Great Plains, but provided no quantitative details. Harper (1955) noted that in many parts of the Barren Grounds there must be as many as ten linear miles of caribou trails to every square mile of territory. He went on to note that *even if* there were only one mile of such trails to each square mile, the total on the Barrens of Keewatin and Mackenzie, Northwest Territories, would equal or exceed all the railway mileage in the United States!

Other studies of the direct effects of trampling exist in association with domesticated livestock, but even these are largely focused on recovery period from trampling, or are "agenda-driven" in an attempt to illustrate the benign influence of grazing (Buckhouse et al. 1981; Kondolf 1993). Moles (1992) described a network of paths produced by trampling by cattle and goats in Burren National Park, Ireland, and drew the interesting conclusion that paths formed by animal trampling develop differently from those created by human trampling. His work suggested that soil depth and slope do not significantly influence the degree of trampling damage by animals, whereas studies on human trampling in path formation have found slope angle a particularly powerful influence on the degree of damage.

Figure 6.2. Closeup of a site near alpine tree line trampled by grazing mountain goats (*Oreamnos americanus*); Hidden Pass, Glacier National Park, Montana.

(a)

(b)

Figure 6.3. (a) Wild-animal trails are visible on a talus slope in the headwaters of Ole Creek, Glacier National Park, Montana. (b) An elephant trail in the Kalahari. Photo (b) courtesy Gary Haynes, from Haynes (1991).

Figure 6.4. Prairie grassland habitat of the North American bison (*Bison bison*); a bison is visible at right of center.

In one study of domesticated livestock (sheep), the direct geomorphic effects of trampling associated with grazing were clearly illustrated. Higgins (1982) sought to determine if animal grazing could produce "terracettes" (Fig. 6.5), a relatively common form of stepped or tiered microrelief found on moderately steep grassy slopes around the world. Many geomorphologists had previously attributed terracette formation to periglacial processes or shallow landslips (Rahm 1962; Higgins 1982, p. 461). Higgins observed the effects of several weeks of grazing in a previously ungrazed enclosure on a moderately steep slope in California. Six weeks of grazing by sheep on a slope with no prior terracette microrelief led to the formation of well-defined treads 20–320 cm wide. These were bare of vegetation and paralleled the slope contour for tens to hundreds of meters. Approximately two weeks of grazing by a mere dozen cattle also produced well-defined grazing steps on a previously smooth west-facing slope (Higgins 1982). Higgins concluded that sheep, cattle, and (based on anecdotal information) horses can indeed produce terracette microrelief. He was not, however, confident in predicting on the sole basis of tread width *what kind* of animal produced grazing terracettes.

In natural areas unaffected by grazing, the existence of terracettes may then offer evidence of the geomorphic influence of natural populations of grazing animals. In areas where both native and livestock animals are known to graze, distinguishing which animals are responsible for terracette formation could theoretically be based on slope steepness, because in general cattle and horses make less use of steeper slopes than do their native counterparts such as deer, elk, and bighorn sheep (Ganskopp and Vavra 1987). Sheep, however, may also utilize steeper slopes, obscuring any such conclusion in an environment where sheep have grazed.

Another example of the direct effects of introduced livestock was presented by Perez (1992b, 1993), who illustrated that cattle may enhance alpine turf exfoliation by the kicking actions of their hooves. Perez (1992b) also summarized the literature dealing with grazing as an agent of turf exfoliation. Perez further noted (1992a) that the reduction in foliage caused by cattle overgrazing in his alpine study area could lead to a deeper penetration of the freezing plane into the ground, in turn producing more frequent needle-ice growth and resulting in higher erosion rates. Higher needle-ice frequency attributable to cattle overgrazing would also inhibit subsequent reestablishment of the native vegetation (Perez 1992a).

In most cases, examination of the indirect geomorphic effects of grazing animals has been restricted to domesticated livestock rather than to free-ranging natural animal populations. (See Simoons [1974] for a good review

Figure 6.5. Terracettes associated with domestic sheep grazing in the Lemhi Range of east-central Idaho.

of some of these impacts, particularly with reference to goats in the Mediterranean region.) Accordingly, geomorphologists have found it necessary to draw analogies from these studies of livestock. A few examples associated with natural populations do, however, exist. Mahaney (1986) and Mahaney and Boyer (1986) examined the behavior of rock hyrax (*Procavia johnstoni mackinderi*) and groove-toothed rats (*Otomys orestes orestes*) in the Mount Kenya afroalpine environment. These small herbivores crop the surface of vegetation until it is bare, leaving it susceptible to the effects of frost heaving, solifluction, and patterned-ground formation (soil stripes) by allowing deeper penetration of the diurnal freezing plane (Mahaney and Boyer 1986, p. 259). More frequent needle ice and enhanced eolian erosion resulting in deflation hollows were the net results – processes and results later described by Perez (1992a) in association with cattle.

Lock (1972) and Pastor et al. (1988) described the soil changes and compaction resulting from trampling by hippopotamuses and moose (*Alces alces*), respectively. Lock noted that the soil compaction prevented rapid infiltration of rainwater, enhancing overland flow and surface erosion. He also noted that the hippopotamus foottrails frequently acted as water channels during rainstorms, becoming slowly incised.

Overgrazing by horses and yaks in the Dalijia Mountains of northwestern China affected the thin mountain soils (Mahaney and Linyuan 1991). Overgrazing led to the degradation of the soil and the development of a massive structure and an increased bulk density from compaction. These conditions in turn induced excessive surface runoff and rill erosion. Forms of mass wasting including solifluction, debris flows, and earth slips were also attributed to the removal of woody shrubs and overgrazing of the alpine grassland there.

Increases in soil bulk density due to trampling and overgrazing have been experimentally described and confirmed for a number of domesticated animals, but particularly for cattle, probably because of their greater size and potential trampling capabilities in comparison to animals like sheep. In experiments in New Mexico, an ungrazed soil with a bulk density of 1.09 g cm^{-1} experienced a bulk density increase to 1.54 g cm^{-1} after light cattle grazing, and up to 1.90 g cm^{-1} on heavily grazed sites (Wood, Wood, and Trombie 1987; Weltz, Wood, and Parker 1989). Weltz et al. (1989) showed that soil bulk density increased by as much as 0.13 g cm^{-1} after only a month of grazing and trampling by cattle. In the same experiment, water infiltration decreased by as much as 75%, and sediment yields in surface runoff increased by as much as a factor of 13. These results clearly illustrate the indirect triggering effect whereby trampling increases bulk density (see Tollner, Calvert, and Langdale [1990] and references therein for more details on livestock trampling), in turn leading to decreased water infiltration and an exacerbation of fluvial erosion (cf. Mahaney and Linyuan 1991). Other studies document the longevity of these effects by examining the recovery time necessary for soils to reach pretrampling conditions (cf. Knapp 1989; Kondolf 1993). Again, however, these studies have focused almost exclusively on domestic livestock rather than free-ranging, native animal populations.

In studies that examine free-ranging animal populations, the focus has been on introduced or feral species, and has primarily been on biogeographic rather than geomorphic implications. Certainly the devastation of both native plants and animals by introduced rabbits in Australia stands as the classic case. Veblen et al. (1992) described the effects of introduced cattle and deer along a vegetational gradient in Argentina from rain forest through xeric woodlands to Patagonian steppe. Although their results primarily focused on impacts to vegetation, soil compaction and subsequent geomorphic modifications must also follow. Another such example is the work of Johnston, Pastor, and Naiman (1993), who examined the influence of moose in northern Minnesota. Ouellet, Heard, and Boutin (1993) illustrated that introduced caribou on Southampton Island, Northwest Territories, Canada, could seriously overgraze the delicate tundra environment and induce environmental degradation within as little as two years. Tisdell (1982) described the widespread disturbances attributed to feral pigs in Australia, which geomorphically "damaged" the landscape by rooting, trampling, and wallowing. Also in Australia, Bowman and Panton (1991) described the trampling effects associated with introduced banteng (*Bos javanicus*) and pigs (*Sus scrofa*) in the Cobourg Peninsula of northern Australia. Pigs rooted extensively in swamps, and around billabongs during the dry season. Banteng trampling was concentrated in areas with monsoon forest and coastal plain vegetation. Associated effects included the knocking over of small monsoon forest saplings (faunalturbation), and trampling of small mammal burrows.

Finally, I must comment on the "trampling" effects of bats. King-Webster and Kenny (1958) described large roof excavations in caves, which they attributed to the roosting effects of bats. As pointed out by Hooper (1958), however, the features described by King-Webster and Kenny are common features in phreatic caves, that is, caves formed beneath the water table under phreatic pressure. The features were not, therefore, the product of a sort of "inverted trampling" (my term, not Hooper's) by bats, although Hooper (1958, p. 1464) did acknowledge that the claws of bats can mark the roof of a cave, "at least to the extent of scratching off powdery surface deposits from the limestone." Hooper felt, however, and I concur, that bat "inverted trampling" represented only a secondary and very minor form of erosion.

Wallowing

Wallowing in either mud or dust is a common activity among mammals. The activity is associated with grooming, heat reduction, and an attempt to achieve relief from insects and parasites (Sowls 1984; Owen-Smith 1988). It has also been reported from oceanic environments, where beluga whales use gravel beds in shallow lagoons to rub off loose epidermis (Frost, Lowry, and Carroll 1993). In doing so, belugas geomorphically mix and stir the bottom gravels and create muddy plumes in the water.

Wallowing is especially notable among pigs and piglike animals (Sowls 1984), as well as among large megaherbivores with relatively hairless skin, including elephants, rhinos, hippos, African buffalo, water buffalo, and warthogs (Owen-Smith 1988). A variety of large seals also wallow extensively (Hall and Williams 1981). Even small rodents frequently wallow in dust in attempting to achieve relief from parasites (Manville 1959). The American bison, or buffalo (*Bison bison*), created extensive wallows across the North American prairie, wallowing in both mud and especially dust (Roe 1970). Large, hairy ungulates such as elk, deer, moose, and the American bison also wallow during their rutting season, and although the reasons for this wallowing are behaviorally quite different, the net effects are of similar geomorphic importance to wallowing associated with grooming.

Quantitative data on the geomorphic effects of wallowing are limited. Manville (1959) reported that Columbian ground squirrels were energetically dusting themselves in dry soil beneath a low tree at alpine tree line. The dusting activity had worn a depression more than 2.5 cm deep into the soil there.

North American ungulates such as the moose (*Alces alces*) and the American elk, or wapiti (*Cervus canadensis*), use their forehooves and antlers to excavate large wallow pits during the autumn rutting season. Van Wormer (1972) reported that male moose produce new wallows each season, and described a typical wallow as about 1.2 m long, 60–90 cm wide, and 5–7 cm deep. Such a wallow would require ~0.04–0.08 m^3 of sediment excavation. Murie (1951) described similar wallows of slightly smaller size used by American elk, but he did not know how often they were used or how many bull elk utilized them.

The propensity of swine to wallow in mud is well known; so well known, in fact, that Mima mounds (see Chapter 7) have mistakenly been called "hogwallows" or "hogwallow mounds" (Arkley and Brown 1954; Cox 1984b), with origins incorrectly attributed to the wallowing action of hogs. True hog wallows have been studied quantitatively by Belden and Pelton

(1976) in the Great Smoky Mountains of eastern North America. Wallows were located in muddy depressions along or near foottrails used by humans, or around small streams. The average wallow was 1.3 × 1.0 m and 25 cm deep; thus, each wallow required the removal of ~0.325 m³ of sediment. All told, forty-eight such wallows were encountered in a variety of elevational zones and during different seasons.

Wallowing effects of the American bison (*Bison bison*) are surprisingly understudied, given their enormous number and widespread nature prior to the coming of European settlement to North America. In the early nineteenth century, there were perhaps forty million bison roaming the American prairie, with additional numbers in forested enclaves in the eastern and southern states (Mielke 1977; Jones and Hanson 1985). Most prairie wallows were apparently *not* associated with salt licks, as was common in the south at such locations as the Great Buffalo Lick in Georgia's Oglethorpe County (~60 km northeast of Athens, Georgia). In the prairie and Great Plains region, the base-rich soils freed the bison to roam great distances rather than being tied to a local salt source. Wallows were excavated by pawing up the earth and subsequently rolling in the dust or mud exposed (Roe 1970).

Whicker and Detling (1988b), in a study examining the interaction of American bison, prairie dogs (*Cynomys* spp.), and pronghorn antelope (*Antilocapra americana*), examined buffalo wallows in a prairie dog "town" in Wind Cave National Park, South Dakota. They simply described "many bare areas, wallows, several meters in diameter" (p. 314), and also noted the presence of heavily used trails and other trampled areas where soil was compacted. Roe (1970), in summarizing literature from the nineteenth century, described wallows 15 or 20 ft in diameter, and 2 ft in depth. Given the vagaries of these early descriptions, I am not comfortable calculating actual volumes of sediment removed, but suffice it to say that it would have been impressive. Such excavated sediment would quickly be fodder for the drying winds of the plains, such that its export from the local wallow site was ensured. Adding to the export of sediment was the dried and caked mud and dust frequently observed on bison by early plains travelers.

How many buffalo wallows were created before the near-extermination of the American bison in the late 1800s? No one knows. Roe (1970) points out that nearly every natural depression in the American and Canadian Great Plains has at one time or another been referred to, sometimes in jest and sometimes in all seriousness, as a "buffalo wallow." It seems clear, however, that thousands of tons of sediment were involved over a vast geographic region. The abrupt cessation of the geomorphic impact of bison on the Great Plains, due to their near extinction via overkilling in under a century, must stand as one of the most striking examples of the creation of a disequilibrium landscape anywhere on earth.

The predominantly seagoing pinnipeds (the "seal" family) seasonally become land dwellers for the purpose of molting, mating, and rearing of newborns. Because pinnipeds form vast colonies, particularly on subantarctic islands (and in many cases, recent studies have seen actual population increases; Boyd 1989), their geomorphic impacts can be substantial. As part of a larger study of the erosional effects of animals at subantarctic Marion Island, Hall and Williams (1981) described the erosional effects of elephant seals (*Mirounga leonina*). During their annual molt, approximately forty-five hundred elephant seals haul their substantial bulk (females weigh about 900 kg, and males up to 3,600 kg; Hall and Williams 1981) inland for distances of up to several hundred meters.

As seals move inland to molt, generally in the same localities used year after year, their great bulk flattens and damages the vegetation. The depressions they create in the molting areas often act as drainage lines which concentrate run-off and initiate small-scale fluvial erosion and transport. Within the molting area they create hollows which, through time, become deepened into wallows. Typical wallows are elongate with rounded ends and an outward bulge in the middle. Measurement of 12 such wallows indicated an average of 3.07 m^3 of material displaced per unit. This displacement may, in part, be explained by compaction although the effect of dilatancy on the granular mass is, as yet, unknown and may inhibit the amount of compaction which is intuitively thought to take place. (Hall and Williams 1981, p. 21)

The wallows also act as catchment sites for water and manure, so that as a seal enters or moves in the wallow, an overland flow of sediment and manure in suspension transports sediment from the wallow downslope toward the coastline. Wallows may coalesce where many seals congregate, forming a "compound wallow," varying from a combination of two to "many tens" of single wallows (Hall and Williams 1981, p. 21). Two such compound wallows had impressive volumes of sediment removed, 768 m^3 and 129 m^3. The net effect of wallows, coalesced wallows, and interspersed vegetated ridges and humps is a very distinctive, hummocky topography. Other studies of wallowing by pinnipeds were carried out by Smith (1988) and Boyd (1989). The latter study focused primarily on the density of breeding Antarctic fur seals (*Arctocephalus gazella*) at Bird Island, South Georgia, but Boyd (1989, p. 180) noted that areas of tussock grassland had been eroded "to various degrees" and "in many areas close to the beaches it had turned to mud." Approximately thirty thousand fur seals were counted during a breeding season at Bird Island. Smith's (1988) study also examined Antarctic fur seals, in his case on Signy Island in the South Orkney Islands. His study primarily detailed the ecological devastation wrought by an expanding population of fur seals (from a few dozen in the 1950s to over thirteen thousand in 1987). On parts of the island over 75% of the vegetation had been either completely destroyed or was in various states of deterioration and removal; turf banks were being destroyed, and soils were being disrupted and subjected to fecal nitrogen enrichment and associated acidification.

Certainly the greatest amount of research on the effects of wallowing has been carried out in association with waterholes and large ungulates in Africa (Flint and Bond 1968; Weir 1969, 1972; Ayeni 1977; Goudie and Thomas 1985; Thomas 1988). The wallowing action is associated with geophagy and the production of pans or waterholes. Elephants especially, but also zebra, wildebeest, hartebeest, and rhinoceros (Flint and Bond 1968; Thomas 1988), excavate and consume saline-rich soils. The excavations in turn enable pond water; this is drunk by a range and number of animals, whose trampling creates an impermeable clay seal and subsequent increased water retention (Goudie and Thomas 1985; Thomas 1988). The focal points of these excavations are frequently large, salt-rich termitaria (Weir 1969, 1972; Thomas 1988), but whatever the starting point, large natural claypan waterholes are the result. These waterholes are subsequently used for wallowing in mud by elephants and other large animals. Flint and Bond (1968) estimated that each individual elephant erodes 0.3–1.0 m^3 of sediment every time such a mud wallowing is undertaken. After the elephants leave the wallow site, the dried and caked-on sediment is carried out of the pan, further enlarging the waterhole. Pans or waterholes thus created by geophagy, drinking, and wallowing range in size from 0.45 to 0.59 km^2 in Kenya (Ayeni 1977); those reported by Thomas (1988) from Zimbabwe were not typically larger than 200 m across. Although deflation during the dry season contributes to the creation of these large pans (Flint and Bond 1968; Goudie and Thomas 1985), their origin is primarily and largely a result of wallowing and the related processes of trampling, drinking, and geophagy.

Geophagy

Geophagy literally is "earth eating," the deliberate consumption of earth (Vermeer and Frate 1975). Unlike lithophagy (rock eating) in reptiles and some birds, geophagy is undertaken to ingest deliberately nutrients or elements that are required by animals but at least temporarily (typically seasonally) unavailable in sufficient quantities (Hebert and Cowan 1971). Geophagy includes actual excavation and consumption of earth material (Cowan and Brink 1949; Calef and Lortie 1975; Holl and Bleich 1987; Mahaney 1987), as well as consumption via "licking" of mineral-rich soils and earth (Cowan and Brink 1949; Stockstad, Morris, and Lory 1953; Hebert and Cowan 1971; Fraser and Reardon 1980; Fraser and Hristienko 1981; Tankersley and Gasaway 1983; Klein and Thing 1989; Moe 1993). At least one case exists where porcupines apparently gnaw and score rock outcrops in order to hone incisors (Gow 1992). Ingestion of sediment by benthic-feeding animals (cf. Nelson and Johnson 1987; Nelson et al. 1987; Obst and Hunt 1990; Ezzell 1991; Avery and Hawkinson 1992) is not here considered to be a form of geophagy, and has been examined in Chapter 5.

Excavation of earth for consumption is carried out via "pawing" the earth (Calef and Lortie 1975; Perez 1993), by use of horns, antlers, or tusks as digging tools (Weir 1972; Mahaney 1987, 1990), by breaking off pieces of termite mounds (Weir 1969, 1972; Ayeni 1977; Davies and Baillie 1988; Mahaney, Watts, and Hancock 1990; Heymann and Hartmann 1991; Izawa 1993), by simple excavation (Mahaney et al. 1990), or by actual biting of the soil or earth material (Oates 1978; Klein and Thing 1989; Heymann and Hartmann 1991). Wallowing is also a common behavior at lick sites, but it is not clear if the purpose of this action is to loosen earth material for ingestion (Perez 1993).

The literature on geophagy among mammalian herbivores, especially ungulates, is diverse, and has been summarized by Jones and Hanson (1985). Geophagy has also been reported for elephants (Weir 1969, 1972) and primates, including a variety of lemurs, monkeys, chimpanzees, and apes (Oates 1978; Ganzhorn 1987; Davies and Baillie 1988; Mahaney et al. 1990; Heymann and Hartmann 1991; Izawa 1993; Mahaney, Hancock, and Inoue 1993).

Actual accounts of the amount of material excavated (i.e., erosion by geophagy) are difficult to obtain and separate from the effects of trampling and soil churning often associated with such sites. At some waterholes and salt licks in southern Africa, elephants actually uproot trees (Coetzee et al. 1979). Numerous other authors have also commented on erosion at wallows and along animal trails that lead to geophagy sites, again emphasizing the interaction of the processes of trampling, wallowing, and geophagy (Eaton 1917; Pack 1928; Cowan and Brink 1949; Murie 1951; Knight and Mudge 1967; Weir 1969, 1972; Skipworth 1974; Calef and Lortie 1975; Singer 1975; Fraser and Reardon 1980; Fraser and Hristienko 1981; Tankersley and Gasaway 1983; Singer and Doherty 1985a; Butler 1993). Phrases such as "well-worn trails," "trails approached the lick from several directions," and "the lick area . . . has been denuded of vegetation and packed hard by . . . trampling" are common in this literature, and indeed these trails are visually striking (see Fig. 6.3a). Weir's (1969) account of trampling and pit excavation at salt licks used by elephants in the Kalahari also includes several striking photographs. The indirect effects of soil trampling and vegetative destruction further exacerbate the erosional situation by reducing infiltration and accelerating surface runoff along animal trails (Laws 1970; Lock 1972; Coetzee et al. 1979). In the following section, I describe the erosion associated with movements of a natural population of mountain goats in their efforts to gain access to a natural salt lick in northwestern Montana.

Trampling erosion associated with a natural salt lick

Mountain goats (*Oreamnos americanus*) are found throughout the U.S. Pacific Northwest, in the Olympic Range of Washington State, the Cascade Range of Washington and Oregon, and in contiguous areas of northwestern Montana, central and northern Idaho, northeastern Oregon, and northeastern Washington (Lyman 1988). Mountain goats of the Olympic Range were artificially introduced in the 1920s; currently, ~1,100–1,200 goats occupy a range of 1,800 km^2 there. Trampling and overgrazing by these introduced goats has "disrupted the floral ecosystem and caused erosion of otherwise stable land surfaces" (Lyman 1988, p. 21).

Unlike the case of the Olympic Range, where goats were artificially introduced, mountain goats also trample and erode surfaces in their natural habitat. The erosional effects of a native population of mountain goats in an American national park where humans have created highway underpasses specifically for goat migration and protection was recently described by Butler (1993, from which this account is excerpted).

Adult mountain goats (Fig. 6.6) are large, hoofed animals, with weights typically ranging from 54 to 73 kg for females, and from 68 to 102 kg for males (Chadwick 1983). Like other ungulates, they require salts in their diet, and thus utilize natural salt licks accessible to them (Jones and Hanson 1985). Erosion occurs as mountain goats develop established trails (Eaton 1917) that lead to natural lick sites. Erosion may also occur along pathways where goats temporarily "bed down" or wallow (Pack 1928).

The Walton Goat Lick is located near the southern tip of, and within, Glacier National Park (GNP), Montana (Figs. 6.7, 6.8). It is located on a cutbank of the Middle Fork of the Flathead River. The mountain goat population in this part of GNP is at least a hundred, with a density (calculated for area of observed use only) of 15.4 goats per square kilometer, five times

Figure 6.6. The North American mountain goat (*Oreamnos americanus*).

Figure 6.7. Mineral lick site (grayish area in center) along Middle Fork of the Flathead River, northwestern Montana, heavily utilized by mountain goats. Several goats are visible on and above the lick site.

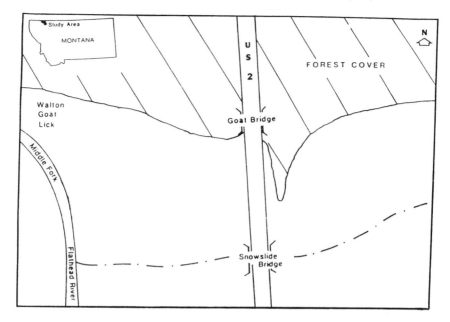

Figure 6.8. Schematic map of a geophagy site heavily utilized by mountain goats along the southern border of Glacier National Park, Montana. The goats formerly approached the lick site through the forested area northeast of the river. Since 1981, they have been directed by fencing and highway underpasses down the ephemeral stream channel and under Goat Bridge or Snowslide Bridge.

higher than any other reported densities in the literature (Singer and Doherty 1985b). These goats make frequent use of the Walton Goat Lick (see Fig. 6.7), but in order to gain access to the lick, they must first negotiate a passage of U.S. Highway 2.

Singer (1975, 1978a) reported on the difficulties encountered by mountain goats as they attempted to cross U.S. 2 and gain access to the lick site. Data from 1975 showed that over a period of ninety days, 328 of 334 crossing attempts occurred over the highway on the way to the lick site, whereas 6 passed underneath a highway bridge spanning a large avalanche path locally called Snowslide Gulch. Singer believed that the dimensions of this underpass were insufficient for attracting the goats – hence the preponderance of surface crossings. On surface crossings, Singer (1978a) reported that the goats preferred to approach the road through forest cover to the north of the Gulch Bridge rather than through the open avalanche-path vegetation characterizing the south-facing slope of Snowslide Gulch (Malanson and Butler 1986). Of the 328 surface crossings, 293 occurred north of Snowslide Gulch along a stretch of highway offering forest cover, and the remaining 35 occurred south of the Snowslide Gulch bridge (Singer 1975, 1978a). Similar figures were obtained on counting return passages from the lick site: Over the same ninety-day period, 345 of 358 passages occurred across the highway, and 13 occurred under the Snowslide Gulch bridge. All but one of the highway crossings took advantage of the forest cover north of the Gulch (Singer 1975, 1978a).

Singer's work (1975) was funded by the U.S. Federal Highway Administration and Glacier National Park, which were planning to upgrade the 1930s-era stretch of highway crossed by mountain goats. Frequent encounters and occasional collisions with automobiles during highway passages led Singer to recommend that goat underpasses be expanded in number and size to accommodate the migrations to and from the Walton Goat Lick.

Highway reconstruction was spurred on 13 February 1979, when a massive snow avalanche completely removed the old highway bridge spanning Snowslide Gulch and deposited it on the floodplain of the Flathead River (Butler and Malanson 1985). This destruction provided the opportunity for beginning the reconstruction of U.S. 2, through the entire goat migration zone, earlier than anticipated. Opened in early 1981, the new section of U.S. 2 (see Fig. 6.8) incorporated a new Snowslide Gulch Bridge (SSGB) with a greatly enlarged goat underpass; a second, new Goat Bridge (GB) with an underpass specifically designed for goat migration; and retaining walls and fences designed to restrict goat access to the surface of U.S. 2 and instead channel the migrating goats through the two underpasses (Singer and Doherty 1985a; Singer 1986; Pedevillano and Wright 1987). All preexisting goat trails leading to highway crossing sites were obliterated (Singer and Doherty 1985a). In steep areas these trails had been entrenched 30–45 cm into the soil (Singer 1975).

Singer and Doherty (1985a) and Singer (1986) described the postreconstruction goat migration patterns from the first season of use, 1981. Over twice as many goat passages downslope (697) were recorded than had occurred in 1975, and 488 successful return passages were also observed. Of the passages downslope to the lick site, 283 utilized the new Goat Bridge underpass, 3 crossed the surface of the highway south of SSGB, and fully 411 used the newly broadened underpass under the new SSGB. Return trips saw 259 passages under GB, 227 under SSGB, and 1 each across the highway north of GB and south of SSGB. These data represent a dramatic increase in the number of goats passing through the low herbaceous and shrubby vegetation on the south-facing slope of Snowslide Gulch, and illustrate that forest cover, although desirable for camouflage prior to a surface crossing, was not deemed necessary for utilizing the broad goat underpass.

Similar postreconstruction migration patterns were recorded for May–August 1984 by Pedevillano and Wright (1987), whose data were not broken down into downslope versus upslope trips. In that study, SSGB had statistically significantly higher rates of crossing than did GB (paired t-test, $t =$ 3.03, $p < .05$). A total of 699 goat crossings were recorded under SSGB, and 300 for GB.

I first noticed goat trails associated with movements of mountain goats to and from the Walton Goat Lick (Fig. 6.9), by way of the SSGB underpass, in 1983 (two years after highway reconstruction); erosion probably began during 1981 and continued through 1982. Because on-site measurements of erosion and sediment movement in Snowslide Gulch would disrupt goat migration and interfere with tourist enjoyment of the goats in their "natural" setting, I used repeat ground photographs to document landscape changes associated with goat erosion between 1983 and 1991. Recognizable rock outcrops and trees provided "benchmarks" in Snowslide Gulch from which erosion effects could be determined. Photographs from 1983, 1987, 1988, 1990, and 1991 provide the framework for the following discussion; however, only those from 1983, 1987, and 1991 are reproduced here.

By July 1983, goat movement along the south-facing slope of Snowslide Gulch had produced a nearly barren patch of exposed sediment to the right of outcrop A on Figure 6.9, as well as an established trail below (and out of sight of) outcrops A and B. An incipient linear scar (C on Figs. 6.10a,b) was visible, with an estimated width of ~1 m and ~15–20 cm depth of incision. In 1983, this scar did not extend as far downslope as outcrop B.

By 1987 (Fig. 6.10a), a large bare patch had developed immediately upslope of outcrop A, indicating serious erosional deterioration since 1983. Even more apparent is the extension of linear scar C to a point well downslope of outcrop B and nearly reaching between outcrops E and F, and the development of a new, major linear scar D. Although it had not existed in 1983, by 1987 (Fig. 6.10a) scar D had reached the level downslope of tree G, and small amounts of sediment had washed beyond G toward tree H. Numerous goat paths can be seen in Figure 6.10a cutting diagonally across the south-facing slope of Snowslide Gulch. Some patchiness of the vegetation between outcrops A and B is also apparent.

Conditions in 1988, although not illustrated here, reflected a continuation of the erosional trends of 1987. The vegetation between outcrops A and B had become sufficiently disrupted to expose bare ground, and scars C and D were somewhat more deeply incised.

By 1990, erosion had nearly reached the level illustrated in the 1991 photo (Fig. 6.10b). In 1990, the erosional patch upslope of outcrop A was com-

Figure 6.9. Herd of mountain goats on the south-facing slope above Snowslide Bridge. Note the heavily trampled appearance and areas of bare surface where turf has been completely removed. (Labels are explained in the text.)

pletely barren of vegetation, a second distinct bare patch had developed below outcrop B, linear scar C extended downslope nearly into the dual rock
outcrops E and F, and scar D almost reached tree H.

Figure 6.10b illustrates the conditions in Snowslide Gulch in 1991, the
eleventh season in which mountain goats heavily used the SSGB underpass.
Linear scars developed beneath outcrops A and B from the bare patches described earlier. Linear scar C extended into the gap between outcrops E and
F, scar D had passed tree H, and a wedge of sediment actually reached the
intermittent stream channel at the base of the slope. In addition, a diagonal
goat trail from A to beyond D, only suggested by vegetation patchiness in
Figure 6.10a, had become well established and extended far beyond D (Fig.
6.10b). The vegetation below letters G-E-F was very patchy; when viewed
in color, it is apparent that major barren areas of exposed soil exist, boding
ill for the future of that portion of the slope.

The total volume of goat-induced sediment erosion since 1981 on the
south-facing slope of Snowslide Gulch was initially estimated for the linear
scars beneath outcrops A and B (hereafter referred to simply as scars A and
B) and for linear scars C and D. Erosion volume was estimated through bi-
noculars, using mature mountain goats for scale. Estimates are, therefore,
accurate only at a gross level, and because only the linear scars are examined the volume data underestimate the total amount of sediment removed
from the slope during the period 1981–91. The estimates do, however, provide a rudimentary indication of the amount and rate of erosion that can be
induced by a concentrated population of mountain goats.

Scar D is by far the largest source of sediment on the south face of Snowslide Gulch. At approximately 20 m long, 1.5 m wide, and 50 cm deep, at
least 15 m^3 of sediment have been eroded due to the goats. Scar C (~12 m ×
1 m × 0.5 m) has lost 6 m^3 of sediment. Scars A (2.5 m^3) and B (2.25 m^3)
are approximately equal in size. The four scars collectively account for
25.7 m^3 of sediment loss, and additional noncalculated losses from goat
trails and smaller patches would probably push the total up to ~30 m^3. Over
the period 1981–91, that is an annual sediment-loss rate of 2.7 m^3. Projected
over a thousand-year time span (and the present animal population), erosion
would allow for 2,700 m^3 of sediment removal caused by concentrating
mountain goats along a migratory path to and from the SSGB. This value is
more than a factor of 10 greater than current rates associated with snow
avalanches in the same environment (Butler and Malanson 1990). The erosion rate on avalanche paths in Glacier National Park is 1,000 B (where
1 B[ubnoff] equals 1.0 mm of ground loss and slope retreat per thousand
years, equivalent to 1.0 m^3 km^{-2} yr^{-1}; Young and Saunders 1986). Rates of

Figure 6.10. (a) Erosion on the south-facing slope above Snowslide Bridge, 22 June 1987, associated with goat migration to the Walton Goat Lick. (b) Continuing deterioration from goat trampling is evident in this view of the south-facing slope above Snowslide Bridge, 22 August 1991. (Labels are explained in the text.)

Figure 6.11. Winter view of the south-facing slope above Snowslide Bridge shows that animals do utilize the pathway during all seasons, but the snowcover must severely restrict any direct erosion common during spring, summer, and autumn; 17 January 1993.

mountain-goat-induced erosion in Snowslide Gulch alone therefore exceed 10,000 B, and this erosion is accomplished primarily in the spring and summer seasons (Fig. 6.11). Additional unknown quantities of sediment may be eroded in association with the passage of goats through the coniferous forest leading to the GB underpass; however, this erosion had already existed at even greater levels during the pre-1981 era as larger numbers of goats passed through the forest before crossing over the highway.

Conclusions

The geomorphic effects associated with mammalian trampling, wallowing, and geophagy are geographically widespread and geomorphically significant. They are, unfortunately, also frequently unquantified. Each process accounts for large amounts of sediment export and transport, sometimes single-handedly, but frequently in a complex interweave of trampling, wallowing, and geophagic activity. The research in this area has been biased toward large mammals and those that form herds or groups. This may be appropriate given the trampling and compacting abilities of large mammals such as elephants and American bison, but their currently restricted geographic ranges suggest that more attention could be paid to the effects of smaller mammals such as rodents: Singularly, they may not have a significant trampling impact, but collectively and through time they also develop distinctive trails and wallows, and must also visit salt licks. Aside from a few casual references to "small animal trails" and beaver runs (cf. Orr 1977; Huntly 1987; Butler 1991b), this area of mammalian trampling influences is virtually unexplored and offers great potential for future research.

7

The geomorphic effects of mammalian burrowing

Introduction

As I have already shown in Chapters 2–4, burrowing serves a diversity of functions, including denning and rearing of young, socialization, shelter from predators and protection from climatic stress, caching of food (Chapter 5), access to below-ground food sources, and sites for seasonal hibernation/estivation (Collias and Collias 1976; Von Frisch 1983; Hansell 1984, 1993; Andersen 1987; A. Meadows 1991; P.S. Meadows 1991; Meadows and Meadows 1991b). In the broadest sense, burrowing encompasses features ranging from shallow daybeds to extensive tunnel complexes. The process of burrowing directly affects geomorphic processes because it bioturbates surface and subsurface sediment, leads to sediment deposition in the form of surface spoil heaps, and may create distinctive landforms. Indirectly, burrowing produces profound geomorphic results in fashions similar to those already examined for other animals: through its influence on soil texture and structure, fertility, and infiltration capacity (cf. Abaturov 1972; Hirsch, Stubbendieck, and Case 1984; Tadzhiyev and Odinoshoyev 1987; Dmitriyev 1989; Dmitriyev, Khudyakov, and Galsan 1989), and the resulting changes both in production of vegetation cover (Del Moral 1984; Gessaman and MacMahon 1984; Huntly 1987; Moorhead, Fisher, and Whitford 1988; Minta and Clark 1989; Peart 1989; Mun and Whitford 1990), and surface runoff and erosion (Turcek 1963; Ursic and Esher 1988; Bykov and Sapanov 1989; Laundré 1993). This chapter examines the direct and indirect geomorphic effects brought about by burrowing. Special emphasis is placed on describing the landforms created by burrowing, and on quantifying the amounts of sediment displaced and entered into hillslope debris cascades.

Burrowing and denning

Burrowing mammals are present on all continents of our planet. "Fossorial" (i.e., adapted to burrowing) rodents, for example, are found on all continents except Australia (Andersen 1987), which is home to numerous burrowing monotremes and marsupials. Burrowing results in surface disruptions and deposition of spoil heaps, frequently at scales sufficiently large to be discernible on small-scale aerial photographs and even on satellite imagery (Vitek 1978; Löffler and Margules 1980; Reading et al. 1989; Rumyantsev 1989; Luse and Wilds 1992).

Burrowing is a process whose origin in mammals dates at least to the Pliocene, the time of appearance of ground squirrels (Dmitriyev 1989). Burrows range in degree of construction along a continuum from simple daybeds to elaborate underground complexes. Mysterud (1983, p. 219) points out that construction of dens for winter-season shelter "is here regarded as no more than a winter bedding," and that "[v]arying degrees of bed construction," ranging from surface scrapes to elaborate burrows, "should simply be viewed as behavioral elements in the ancient mammal coordination pattern connected with the construction of burrows" (p. 222). In the following sections, I examine burrows of all forms created by mammals, ranging from shelter beds/burrows to those constructed by fossorial herbivores in search of below-ground tubers and other food sources. It would be impractical to describe the burrows of every mammal; rather, emphasis (as in Chapter 5) is placed on diggings for which some quantitative data are available concerning amounts of sediment displaced. I first examine burrowing associated with the more primitive marsupials and monotremes, and follow with the placental mammals.

Burrowing by monotremes and marsupials

The Prototheria, or monotremes, are represented by the Australian echidna (*Tachyglossus aculeatus*), or spiny anteater, found throughout Australia, and by the duck-billed platypus (*Ornithorhynchus anatinus*) of eastern Australia and Tasmania. Both are burrowing mammals, but little is known of the habits of the echidna beyond the fact that it is nocturnal, spending the day underground in excavated burrows or rock cavities (Griffiths 1978). Female echidnas do not excavate nesting burrows. Echidnas can escape from danger by rapidly digging a shallow burrow, leaving only a mound of spines exposed above the surface (Morcombe 1968).

The platypus, perhaps because of its particularly puzzling and bizarre appearance, has been the subject of more intensive investigations. Platypuses excavate both resting burrows and denning burrows. The former may be a simple creation whereby the animal buries itself up to half its body-length in mud and gravel (Burrell 1927), or it can be a more complex excavation. They are created by both males and females at all times of year, and are usually semicircular excavations entered through a tunnel under the roots of large trees (Burrell 1927). Although few data exist on the amount of sediment excavated in association with these resting burrows, Burrell suggests that their diameter is much greater than that of nesting burrows.

Nesting burrows are dug by a single female in riverbanks for the purpose of laying eggs and rearing of the hatched young. Griffiths (1978, p. 218) reports that "the female digs a serpentine burrow between 35 and 50 feet long . . . from the entrance, which is in the side of the bank usually above water level, to a chamber containing a nest of grass, leaves, willow roots etc." The burrows are preferentially excavated into fine-grained alluvium, and sandy banks are specifically avoided because of the likelihood of tunnel collapse (Burrell 1927). Tunnels up to 100 ft (~30 m) in length have been reported, and Burrell (1927) noted that little actual deposition of material from the tunnels onto the surface occurs. Instead, the platypus compacts the excavation with its body and tail so that the excavated material occupies only about one-half its original volume. Certainly this must have major ramifications on reducing the ability of rainwater to infiltrate into ground surfaces where platypus nest burrows occur, but I know of no studies that have examined this aspect of their burrow excavations.

A few species of marsupials (Metatheria) are found in North, Central, and South America, but the vast majority (about three hundred different species) live in the Australasian region. Burrowing marsupials of Australia include the rabbit-eared bandicoot, or bilby (*Thalacomys lagotis*), which lives in woodland and open-steppe country; the small Australian marsupial mole (*Notoryctes typhlops*), living underground in semidesert sand-dune country; and the wombats, which are efficient burrowers, digging with their forepaws and pushing soil away with their feet (Morcombe 1968).

Burrowing among the marsupials is apparently most highly adapted in the hairy-nosed wombat (*Lasiorhinus latifrons*) of Australia (Löffler and Margules 1980). Its strong claws and powerful legs make it capable of burrowing not only through compact soil, but also through thick layers of calcrete. A wombat burrow hole is about 0.5 m in diameter, whereas a wombat warren is comprised of several burrows and forms a distinct mound rising 0.5–1.0 m above the surrounding topography (Löffler and Margules 1980). Vegetation surrounding warrens is typically completely denuded, creating large circular areas of bare ground with warrens in the center of such a size that they are visible on Landsat digital satellite imagery. Warrens may have a diameter of 20–30 m, and on average have ten to fifteen burrows, and a wombat colony typically has between ten and thirty warrens covering an area as large as 1 km² (Löffler and Margules 1980). Unfortunately, depth information associated with wombat burrowing was not presented by Löffler and Margules, rendering quantitative estimates of sediment displacement tenuous at best. Nevertheless, from their published descriptions, photographs, and satellite imagery, it is emphatically apparent that wombats are capable of landform creation and massive sediment displacement. Such activities in turn must have major, but completely unanalyzed, influences on rainfall infiltration, splash erosion, and surface runoff.

Large herbivores

Large herbivorous mammals, including the ungulates, do not as a rule create burrows. Their propensity to wallow has already been discussed in Chapter 6. In some cases, however, large ungulates are known to excavate daybeds, which are, as pointed out by Mysterud (1983), the simplest forms on a continuum of burrow architecture. Daybeds are usually located in topographic hollows, where only minor excavation is required. Bighorn sheep, for example, typically scratch on the bed with a foreleg, removing rocks and pebbles that have rolled down the slope and onto the resting site (Geist 1971). Sheep beds are used over and over, and, in concert with the secondary effects of frequent urinations and defecations around the bedding sites, have a pronounced effect on the surrounding vegetation (Fig. 7.1). Vegetation enrichment in association with daybed sites may, therefore, protect the slope and counter any erosional effects induced by daybed construction and areally associated trampling. Deer and American elk (Fig. 7.2) also utilize daybeds, but on slopes of sufficiently low angle that their overall geomorphic influence is questionable.

Carnivores and omnivores

Many carnivores and omnivores burrow for denning purposes as well as for daily or seasonal shelter. The burrowing habits of some carnivores have been noted in the literature, but frequently den-site characteristics are described without any quantitative description of the actual den/burrow size. In the following sections, data are presented on several carnivores and om-

Figure 7.1. Bighorn sheep (*Ovis canadensis*) resting in a daybed in a heavily trampled site near Haystack Butte, Glacier National Park, Montana.

Figure 7.2. American elk (*Cervus canadensis*) daybeds in a subalpine meadow; such daybed sites become heavily trampled. Yellowstone National Park, Wyoming.

nivores, predominantly from North America, for which some quantitative estimates of burrow number and/or sediment displacement are available.

Grizzly bears

The grizzly or brown bear (*Ursus arctos*) is not only a formidable digging machine in search of food (Chapter 5), but is also an effective mechanism for sediment displacement in its construction of daybeds and winter dens. Daybeds are typically found in vegetatively sheltered locations such as tree windfalls, shrub thickets, or forest floors (Craighead and Craighead 1972; Blanchard 1983; Mysterud 1983; Zager, Jonkel, and Habeck 1983). Daybeds are typically located in contour depressions, where a minimal amount of ex-cavation produces a maximum of shelter. Forest duff is scraped away to the mineral soil, which may be excavated to depths of ~0.5–46 cm (Craighead and Craighead 1972; Mysterud 1983). Most beds are shallow, but Craighead and Craighead (1972) reported that some daybeds mimic shallow den bur-rows, reaching depths of >1 m. Mysterud (1983) distinguished among three types of den, based on their morphologic characteristics and purpose:

1. *Hidebeds* are located in dry and sheltered spots under extreme cover.
2. *Watchbeds* are located on topographic sites offering excellent hearing and scent control (grizzly bear eyesight is notoriously weak), such that a bear could lay alert and survey the surroundings; and
3. *Coolbeds* are excavated in wet, sandy soil on shady and well-sheltered sites, presumedly to be used during the peak of summer heat.

During the winter months, typically from late October into if not through April, grizzly bears enter a period of dormancy during which they occupy their winter den. Most grizzlies do not occupy natural cavities in the ground, but instead excavate their dens into the ground surface (Craighead and Craighead 1972; Pearson 1975; Werner 1977; Vroom, Herrero, and Ogilvie 1980; Servheen and Klaver 1983; Interagency Grizzly Bear Committee 1987; Butler 1992). Most dug dens collapse during the period of spring snowmelt and summer rains, so that it is uncommon for dens used during one winter to be reoccupied in subsequent years (Reynolds, Curatolo, and Quimby 1976; Judd, Knight, and Knight 1986). The sediment introduced into hillslope debris cascades by dens excavated on slopes is rapidly re-moved by subsequent erosion; after a period of <75–100 years, slope pro-cesses effectively obliterate evidence of previous denning (Vroom et al. 1980).

In cordilleran locations, where most grizzlies survive in Eurasia and western North America, dens are typically located above or near upper tree line, dug approximately horizontally into slopes averaging 28–35° (Craighead and Craighead 1972; Pearson 1975; Werner 1977; Vroom et al. 1980; Servheen and Klaver 1983; Butler 1992). Even in the relatively mild "rain forest" environment of southeast Alaska and coastal British Columbia, grizzlies den on steep slopes in order to avoid den inundation by runoff; there, however, they typically excavate dens beneath large-diameter Sitka spruce (*Picea sitchensis*) or the base of large tree snags (Schoen et al. 1987). Northern exposures are apparently avoided as den sites, but all other slope aspects may be used, although Vroom et al. (1980) showed that grizzlies in Banff National Park, Alberta, Canada, preferred leeward slopes, away from prevailing westerly winds. Den sites are located such that a deep cover of snow provides insulation for the dormant animal from the extremely cold winter temperatures.

Grizzly dens comprise two morphologic elements: a tunnel leading to the outside, and an interior chamber. The typical tunnel displaces ~1.7 m^3 of sediment onto a spoil heap downslope of the tunnel mouth, and the typical chamber displaces 2.6 m^3. I calculated these figures (Butler 1992) from published values of representative dens in other works (Craighead and Craighead 1972; Werner 1977; Interagency Grizzly Bear Committee 1987), and used them to calculate the amount of sediment displaced annually by the roughly two hundred grizzly bears in Glacier National Park (GNP), Montana. I assumed that, although cubs do not dig separate dens, some adults produce more than one den prior to being satisfied as to structural integrity (Butler 1992). Each of the two hundred dens displaced about 4.3 m^3 of sediment downslope, accounting for an annual sediment removal rate of 860 m^3. Over the past hundred years, during which the GNP area has been protected by the U.S. federal government, this accrues to 86,000 m^3 of sediment removed by grizzly bear den activities in Glacier National Park.

Black bears

Black bears (*Ursus americanus*), which occur throughout much of the forested regions of North America, use both daybeds and winter dens, as do grizzly bears. However, black bear dens – unlike those of the grizzly bear, virtually all of whose winter dens are excavated – vary as to whether they are excavated into soil or into trees and logs (Tietje and Ruff 1980; LeCount 1983; Manville 1987). Black bear daybeds also vary as to the degree of excavation involved. Considerably smaller than grizzlies – adult males (boars) weigh ~200–400 lb, and females (sows) 150–300 lb (Brodrick 1959; Wilkinson 1993) – black bears also lack the grizzly bears' muscular shoulder hump, and possess much shorter claws (which aid in more efficient tree climbing). The geomorphic influence of an individual black bear is thus considerably less than that of a grizzly; but given its much greater modern-day geographic range and larger numbers, the black bear is a widespread agent nevertheless.

Excavation of dens and daybeds into soil, rather than into upright or fallen logs, natural crevices, or simple surface bedding and nesting with excavation, is largely a function of geographic location and the concomitant severity of the winter there. In milder climates such as the southeastern United States, few dens and daybeds are excavated into the surface; rather, daybeds are located in simple, naturally occurring surface depressions, and dens are located in stumps, fallen logs, and in some cases in surface depressions similar to those used for daybeds (Hamilton and Marchinton 1980; Johnson and Pelton 1983; Wathen, Johnson, and Pelton 1986). In locations where daybeds are constructed, beds may vary from almost no excavation to deep, well-excavated bowl-shaped structures (Mollohan 1987). Little quantitative data exist as to the amount of sediment excavated for black bear daybeds, but Johnson and Pelton (1983) provided dimensions of $1.35 \times 0.73 \times 0.23$ m for daybed length, width, and depth, respectively. These data would suggest that each black bear bed displaces ~0.23 m^3 of sediment.

In higher latitudes and altitudes, the greater severity of the winter season induces black bears to excavate winter dens into the soil, where greater insulative capacity exists (frequently in concert with a heavy overlying blanket of snow). The period of hibernation is also controlled primarily by climate. Like grizzly bears, black bears rarely reuse excavated dens, instead typically digging new ones each year (Tietje and Ruff 1980; Klenner and Kroeker 1990).

The structure of an excavated black bear den is quite similar to that of a grizzly bear, with a tunnel leading to the actual den site. Depth beneath the surface varies with slope and aspect, but because the climatic range of the black bear is so much more extensive than the grizzly's, no generalities emerge concerning either of those variables. Although tunnels leading to the interior den site may be as long as 7 m (LeCount 1983), ~1 m in length seems to be the norm (Tietje and Ruff 1980; Beecham, Reynolds, and Hornocker 1983; Kolenosky and Strathearn 1987). The amount of sediment excavated from the tunnel varies by sex and age of the bear involved, but data provided in several papers illustrates that an adult male tunnel displaces ~0.30 m^3, whereas the tunnel of an adult female displaces ~0.25 m^3 of sediment (Tietje and Ruff 1980; Beecham et al. 1983; Kolenosky and Strathearn 1987).

The interior den itself seems to vary in size somewhat more than the tunnel leading from the surface. In Alberta, Canada, interior chambers of adult males displaced 1.17 m^3 of sediment, and those of adult females displaced 0.80 m^3 (Tietje and Ruff 1980). Ross, Hornbeck, and Horejsi (1988) also described an Albertan den, 0.91 m^3 in size. The opening of that individual den had been enlarged by a foraging grizzly bear that had actually killed the black bear within. In west-central Idaho, where Beecham et al. (1983) did not distinguish interior dens by sex of the bear, cavities averaged 0.83 m^3. In Ontario, Canada, solitary adult black bears regardless of sex created den cavities that displaced 0.56 m^3 of sediment (Kolenosky and Strathearn 1987). Finally, at two different sites in Alaska, den chambers undifferentiated by sex of creator varied in size from 0.81 to 1.11 m^3 (Schwartz, Miller, and Franzmann 1987).

Combining the range of 0.25–0.30 m^3 for tunnels with an apparently typical interior den of 0.80 m^3 in size, then, black bear den-and-tunnel complexes each displace ~1.05–1.10 m^3 of sediment, with larger examples not uncommon. As with the grizzly bear, in many cases this sediment is deposited on slopes of at least 20° (Tietje and Ruff 1980), so that virtually all the displaced sediment migrates downslope rather than acting to backfill the den. Each den created is therefore a distinct erosional site, and geographically the geomorphic effects of black bear denning are concentrated at higher altitudes and latitudes, where the more severe winters induce underground denning.

The Canidae family

Wolves, coyotes, dogs, and foxes are members of the Canidae family, many of which illustrate burrowing behavior. Some domesticated dogs are known for their abilities to displace fossorial animals by digging, and my own two domesticated dogs have created several daybed excavations, each about 30 cm × 30 cm × 10 cm in volume in a portion of our family yard. The emphasis to follow, however, is on wild natural populations of Canidae.

Wolves (*Canis lupus*) are the largest wild canines, and they excavate natal dens where the terrain will structurally support tunneling; preferred are sites in sandy, well-drained soils in concert with tree and shrub roots (Heard and Williams 1992). Typical wolf dens have entrances measuring roughly 71 × 43 cm, leading into tunnels about 30 cm in diameter. Spoil mounds outside the den tunnel are about 1.5 m wide by 3 m long (Mech 1966).

Wolves that follow herds of northern caribou have a den-density range of 0.64–1.92 per ten thousand kilometers, with the highest concentration in the tree-line zone, where tree roots provide required support. In such areas, dens are long-lasting and may be reused, whereas those excavated into tundra frequently collapse (Heard and Williams 1992). The low overall number of wolves in northern environments today, in concert with low den densities, must imply that their overall geomorphic effectiveness is spatially insignificant.

Among the foxes, the denning of the arctic fox (*Alopex lagopus*) has been the focus of a number of studies in wildlife ecology that report data on den density and/or site characteristics (Chesemore 1969; Eberhardt, Garrott, and Hanson 1983; Garrott, Eberhardt, and Hanson 1983; Smits, Smith, and Slough 1988; Prestrud 1992). Arctic fox dens are found in appropriate terrain across the Arctic coastal plain of Alaska, Canada, Russia, and several Arctic islands. Foxes den both for shelter and for natal purposes (Eberhardt et al. 1983).

Dens of arctic foxes cover sufficient area that they are easily detected in aerial overflight surveys (Chesemore 1969; Eberhardt et al. 1983; Garrott et al. 1983; Smits et al. 1988). Den sites are spatially limited to areas where permafrost is sufficiently deep to allow burrowing: Where permafrost is within 1 m of the surface, burrows typically collapse and are abandoned (Garrott et al. 1983). Other site characteristics favoring den production include south-facing den burrow openings and well-drained soil conditions (Smits et al. 1988; Smith, Smits, and Slough 1992). Stable den sites are relatively permanent landscape features: Dens have been estimated as being over three centuries old, being enlarged with additional entrances in successive years (Smits et al. 1988).

Primary den characteristics include large numbers of burrow entrances (up to thirty-three per den in the Colville River delta of Alaska; Eberhardt et al. 1983). Den entrances range in general from about 9–35 cm in diameter (Chesemore 1969; Smits et al. 1988). Large, rounded soil mounds represent the spoils excavated from the extensive den system (Garrott et al. 1983) and range up to an average of 4.5 m in height on Herschel Island in the Beaufort Sea off the Yukon coast (Smits et al. 1988). Individual dens cover large surface areas, exceeding 100 m^2 (Chesemore 1969; Smits et al. 1988) and ranging up to >250 m^2 (Eberhardt et al. 1983). Prestrud (1992) mapped "large" natal dens and "small" shelter dens with one to three openings on Svalbard (Spitsbergen), and noted the former outnumbered the latter by more than a 3 : 1 ratio. Dens also impact the surrounding vegetation ecology, with much more lush tundra conditions on den mounds than in the surrounding tundra (Garrott et al. 1983; Smith et al. 1992).

In the process of spoil-mound creation and faunalturbation, arctic foxes displace the normal cryoturbational processes operative in the arctic tundra (Smits et al. 1988). Foxes mix material through burrowing, enriching the organic content of the soil and forming humus-rich surface horizons. Fox den and spoil-mound soils also have consistently higher temperatures than the surrounding undisturbed permafrost (Smits et al. 1988; Smith et al. 1992), but it "is not clear whether den sites are inherently warmer than nearby sites before foxes start constructing the dens, or if the favourable soil thermal regime and the lowering of the permafrost table on den sites result from the presence of burrows, which act as ventilation ducts" (Smits et al. 1988, p. 15). Even in the face of such a chicken-and-egg conundrum, it is clear that arctic foxes geomorphically and vegetatively influence large localized areas in the arctic lowlands. Smith et al. (1992, p. 328) summarized the role of arctic fox burrowing by noting that it "must be considered a vital component in nutrient cycling in tundra ecosystems."

Otters

More than a dozen species of otters exist around the world, but excluding the sea otter (see Chapter 5), the most common are probably the European (*Lutra lutra*) and the Canadian (*Lutra canadensis*) otters (Mason and Macdonald 1986). Otters do not excavate their own dens, but rather use natural formations, human-built structures, and dens built by other animals; active and abandoned beaver bank dens and lodges (see Chapter 8) are especially favored (Melquist and Hornocker 1983). Existing dens and tunnels are enlarged or altered by otters (Melquist and Hornocker 1983), but virtually no quantitative data exist as to the amount of sediment otters move in these modifications.

Badgers

As with the otter, there are several living species of badgers around the world (Long and Killingley 1983; Neal 1986). Certainly the most heavily studied of this group are the European (or Eurasian) badger (*Meles meles*) and the North American badger (*Taxidea taxus*) (Fig. 7.3), although Long and Killingley (1983) provide excellent descriptions of the digging habits of ferret badgers (*Melogale*), ratels or honey badgers (*Mellivora*), hog badgers

Figure 7.3. Badger (*Taxidea taxus*) beneath burrow entrance, Laramie Basin, Wyoming.
Note the large amount of sediment deposited downslope from the den entrance.

(*Arctonyx*), and stink badgers (*Mydaus*). The natural ranges of both the European and North American badgers cover vast areas; European (Eurasian) badgers cover all of Europe south of the tundra, and all of Asia south of the Arctic tundra and north of a line approximating the 35th Parallel. North American badgers are found generally south of the Arctic tundra, with minor numbers only in Canada's forested tracts. The bulk of the population occurs from the Pacific states through the mountains and prairies eastward into Ohio, and southward into Mexico (Long and Killingley 1983). In the case of both Eurasian and North American badgers, human land-use pressures and disturbances have caused population declines and shrinking areal coverages (Moore 1990; Skinner, Skinner, and Harris 1991).

Badgers are known as the fastest diggers of any mammalian species, so fast that "if several men with shovels tried to capture a digging badger, it would be surprising if they could catch up with it" (Moore 1990, pp. 22–4). Badgers dig holes and burrows to capture prey, to escape and avoid danger, to sleep in during the day or night, to sleep in during a period of winter inactivity, and to rear young (Lindzey 1976; Long and Killingley 1983; Neal 1986; Neal and Roper 1991; Roper 1992).

Burrow systems, or "setts," of the North American badger are not as extensive as those of the European badger (Long and Killingley 1983), probably because of the solitary nature of the North American badger (Messick and Hornocker 1981), but they produce much more extensive predatory diggings. Burrows associated with predation (see Chapter 5) are primarily constructed during the summer months (Messick and Hornocker 1981). Badger resting burrows can be extensive, with tunnels of ~20–25 cm in diameter, interior chambers measuring $20 \times 17 \times 25$ cm, and lengths in excess of 2 m (Long and Killingley 1983). Spatially, these burrows are concentrated in areas of high fossorial rodent populations, which serve as the primary food source for the North American badger (Messick and Hornocker 1981). Natal dens are typically twice the size of resting burrows (Lindzey 1976), with correspondingly large spoil mounds (see Fig. 7.3) and tunnels more than 5 m long.

Indirectly, the North American badger affects surface runoff via its burrowing habits. Long and Killingley (1983) emphasized the capability of the badger to facilitate soil development through mixing and aeration (Fig. 7.4), as well as through the addition of organic matter. Soils aerated in such a fashion would, in concert with extensive burrow systems, reduce surface runoff and erosion because of their greater infiltration capacity.

Setts of the European badger are described in detail by Neal (1986), Neal and Roper (1991), and Roper (1992). European badger setts are long-lasting

Figure 7.4. Active and in-filled badger burrows, Lemhi Valley, Idaho. Scale is in inches. Light-colored parent material is volcanic ash, overlain by a rich brown grassland soil. The burrow in ash has been filled in with soil material from above.

features on the landscape, with several instances reported of setts several hundred years in age (Neal 1986; Roper 1992). As with the North American badger, the geomorphic impact of the European badger is minimal in global terms but of great geomorphic and ecological importance locally (Neal and Roper 1991). Roper reports individual setts with over a hundred entrances covering an area of more than a hectare, and tunnel systems and the amount of excavated soil are in excess of their North American counterparts (Neal and Roper 1991; Roper 1992). Locally, setts alter the topography by the widespread presence of spoil heaps containing in some cases "many tonnes" of soil (Neal and Roper 1991), and must also reduce surface runoff in a fashion similar to those of the North American badger.

The only geomorphological study specifically to quantify the burrowing effects of the European badger was carried out on wooded hillslopes in the Belgian Ardennes by Voslamber and Veen (1985). They found that excavated soil located on the surface was primarily associated with badger and rabbit burrowing, and rabbits were only important at two of their eight study sites. The amount of excavated soil per site averaged 7,003.6 kg ha^{-1}. The average length of all recorded badger spoil mounds was 4.32 m. At actual badger mound areas, mass transport equaled 15.05 g cm^{-1} yr^{-1}, but when averaged over the entire study area, the direct contribution to mass transport by badgers was a meager 0.0192 g cm^{-1} yr^{-1} (Voslamber and Veen 1985, p. 82). They concluded that badger mass-transport rates are from one to two orders of magnitude lower than those for seasonal soil creep, and that "[t]he direct geomorphological effect of badger digging in this temperate humid forest environment is, therefore, insignificant" (p. 82).

Armadillos

Approximately twenty species of omnivorous armadillos are found in the New World, but the natural history of only one species, the nine-banded armadillo (*Dasypus novemcinctus*), has been studied in any detail (Smith and Doughty 1984). The recent history of the nine-banded armadillo in the U.S. South is an interesting example of the expansion of the effective range of a geomorphic process. This species extended its range northward from Mexico into the United States as recently as the 1850s (Zimmerman 1990), and is now found in every Gulf State, as well as in Georgia, Arkansas, Oklahoma, and southernmost Kansas (Smith and Doughty 1984). All geomorphic effects of armadillos in the United States are, therefore, <150 years old.

Nine-banded armadillo burrows serve as shelters and natal sites for rearing young. Zimmerman (1990) examined 113 armadillo burrows in central Oklahoma and found that all were characterized by a spoil mound composed of layers in front of the burrow entrance. The bottom layer of the spoil mound was made up of the initial soil excavated; the overlying layers comprised mixtures of leaf fragments, sticks and other organic material, and soil.

Burrow entrances may be oriented toward the south for warming purposes (Smith and Doughty 1984), although Zimmerman (1990) reported orientation data showing no preferred trend. He attributed the absence of a preferred orientation in the Oklahoma burrow entrances to protection provided by perennial vegetation (obviating the need for an alignment designed to maximize warmth), and to burrow collapse resulting from winter freeze–thaw action on southern orientations.

Nine-banded armadillos tend to burrow in sloping terrain (Zimmerman 1990), suggesting that excavated material comprising the spoil mounds is subject to subsequent erosional removal by wind and running water. The mean ground slope for the 113 burrows in Zimmerman's study was 16°, and tunnels sloped downward 29° on average. Burrow entrances in Zimmerman's study were about 22 cm wide and 20 cm high, with tunnel diameters in the 18–20 cm range. No data have been reported in the literature, however, about the overall length of tunnels, from which volumetric data concerning sediment removal could be calculated. Smith and Doughty (1984) did report that tunnel lengths could range from as little as 30 cm, for individual shelter burrows, to over 5 m in association with large natal dens. Natal dens typically have several active entrances, and are about 1–1.5 m below the ground surface (Smith and Doughty 1984).

Rodents and other small herbivorous or insectivorous mammals

In the following sections I examine the geomorphic influence of smaller mammals from the lagomorph, rodent, and insectivore orders. This discussion, like previous ones, is not an exhaustive description of the burrowing activities of all members of these families, but instead concentrates on those animals whose burrowing habits have received broad research attention in the ecological and the geomorphological literature. The smallest mammals, such as shrews (Crowcroft 1957), voles (Martinez Rica and Pardo Ara 1990), and varieties of mice (Ellison 1993) have received very little such attention, and are not described in further detail here. (For good examples of the potential for studying the hydrologic effects of small burrowing mammals, see Ursic and Esher [1988] and Martinez Rica and Pardo Ara [1990];

and for an excellent example of a description of burrow characteristics and volume loss of sediment associated with two small desert rodents in Jordan, see Hatough-Bouran [1990].)

Rabbits

As any child who has read Beatrix Potter's *The Tale of Peter Rabbit* could tell you, rabbits live in holes underneath the roots of large fir trees. As Australians and Europeans can attest, they also dig burrows in a variety of other vegetation types, and their burrows vary greatly in size. (Anthropogenically induced geomorphic change in association with rabbits occurs in Australia, where "warren ripping" is one method of control of the rabbit population; Parker, Myers, and Caskey 1976). Groups of burrows and spoil mounds, or rabbit warrens, may be the dominant local landform in favorable circumstances. In spite of their widespread occurrence and familiarity, however, rabbits have been greatly understudied as geomorphic agents. The research that has been done has focused exclusively on the European rabbit (*Oryctolagus cuniculus* L.); no corresponding research has been carried out on American rabbits or hares.

Characteristics of individual rabbit dens and large warrens have been described from a diversity of habitats in western Europe and Australia (Parker et al. 1976; Kolb 1985, 1991a,b, 1994; Cowan 1991, and references therein). Burrows vary in type and extent from enlarged scrapes to large interconnected structures (Kolb 1985). Density of burrow entrances can be quite high, ranging from 112 (Kolb 1994) to nearly 500 per hectare (Kolb 1991a), and individual warrens frequently have more than thirty entrances (Cowan 1991). Kolb (1994) reported a ratio of burrow entrances to adult rabbits on the order of 9 : 1.

Diameters of warren tunnels fall in the range of 10–15 cm (Kolb 1985; Cowan 1991; Rutin 1992), and tunnels may extend in excess of 2 m (Kolb 1985), but few data concerning the actual amount of sediment removed and subjected to subsequent erosion have been forthcoming. Kolb (1985) did, however, attempt to estimate the relative enclosed volume of burrows. He assumed two different average burrow diameters, 10 and 15 cm. Calculated values for the 10-cm diameter ranged from 0.0187 to 0.0611 m^3 per burrow, and for 15 cm ranged from 0.0421 to 0.1275 m^3 per burrow. Extrapolating those values to Kolb's (1991a, 1994) density values of 112–500 burrow entrances per hectare, one derives a minimal displacement figure of 2.09 m^3 ha^{-1} (0.0187 m^3 per burrow × 112 burrows of 10 cm diam.), and a maximum of 63.75 m^3 ha^{-1} (0.1275 m^3 per burrow × 500 burrows of 15 cm diam.). These values, although only crude extrapolations, illustrate the potential quantitative influence of rabbits on a local environment.

The rabbit warren studies described above were, however, primarily conducted by wildlife managers, not geomorphologists. Only two studies concerning the geomorphic effects of rabbits have been published in the geomorphological literature. Voslamber and Veen (1985) examined the role of both rabbits and badgers on wooded hillslopes in Belgium. As described in the earlier section on badgers, rabbits were less significant than badgers in producing surface sediment. Nevertheless, at one site dominated by rabbits, Voslamber and Veen estimated that 71,308 kg of soil per hectare were produced by rabbit burrowing, although another nearby site yielded only 475 kg ha^{-1}. They concluded that badgers and rabbits may be the dominant process of production of available sediment in the localized areas affected by those animals, but that averaged out over the entire study area, the effects of the animals were one to two orders of magnitude below that of seasonal creep.

The other study of rabbits in the geomorphological literature was by Rutin (1992), who examined the role of rabbits in modifying the surface of sand dunes along the coast of the Netherlands. He reported extensive "caves" dug by rabbits, but his descriptions and photographs reveal that these "caves" are simply the extensive warrens typical of the European rabbit. He noted both indirect and direct geomorphic roles of rabbits: Indirectly, rabbits may increase the potential for erosion by raindrops, wind, and running water by reducing soil cohesion, breaking water-repellent crusts, and by grazing and reducing vegetation cover. More directly, the rabbits caused downslope transport of sediment in the vicinity of the extensive warrens. Rutin also described how warren excavation was responsible for a terrace-step topography, where the sand mounds deposited below warren entrances acted as anchor points for vegetation. Through time, the development of stepped slopes near "cave-hole" entrances resulted. Rutin also correctly pointed out that not all aspects of rabbit burrowing are detrimental and lead to an increased erosion potential, noting that "cave-holes" or burrows improved soil permeability and therefore reduced the likelihood of developing overland flow.

Porcupines

Although porcupine digs associated with foraging for food have received some attention in the geomorphological and ecological literature, very little research has focused on their digging habits associated with den production

(Saltz and Alkon 1992). It is known that the Indian crested porcupine lives in clans that occupy a single den (Saltz and Alkon 1992), but data on the number and size of these dens have not been forthcoming.

Beavers and muskrats

Beavers, mountain beavers, and muskrats excavate bank burrows and dens throughout their respective ranges. The geomorphic effects of true beaver, *Castor canadensis* in North America and *C. fiber* in Europe, are the focus of Chapter 8 and are not discussed further here.

Mountain beavers (*Aplodontia rufa*) live in the American Pacific northwest, from the central Sierra Nevada to southern British Columbia (Orr 1977). They excavate extensive networks of irregular tunnels just below the ground surface, characteristically with many surface openings (Beier 1989; Engeman, Campbell, and Evans 1991). Few other data exist as to the geomorphic effectiveness of the mountain beaver, although Engeman et al. (1991) noted that they keep their tunnels clear by pushing materials out of the openings, that is, they create spoil mounds at the tunnel entrances.

The muskrat (*Ondatra zibethicus*) is an important herbivorous species in marsh environments across North America (Messier and Virgl 1992), currently distributed in every U.S. state and Canadian province except Florida (Farrar 1992). It has also been introduced in Europe, where it has expanded its populations to a level at which it is frequently considered a pest. The muskrat is also known as "mushrat" and in Canada by the Cree name "musquash" (Farrar 1992).

Muskrats occupy two different forms of dwelling: bank burrows and marsh lodges constructed of vegetation (Messier, Virgl, and Marinelli 1990; Wainscott, Bartley, and Kangas 1990; Messier and Virgl 1992). Bank burrows are excavated in shoreline banks in a fashion similar to those of beavers (see Chapter 8), and are apparently the preferred form of habitat for muskrats; aquatic lodges are constructed when population levels exceed a marsh's capability to provide sufficient bank burrows (Messier et al. 1990; Messier and Virgl 1992). Bank burrows occur at densities ranging from about two to three burrow systems per 100 m of shoreline (Messier and Virgl 1992). A tunnel entrance located beneath the marsh low-water line provides access into a tunnel angling upward into the bank burrow (Farrar 1992). Unfortunately, no quantitative data exist as to the amount of sediment displaced by muskrat burrowing, a situation similar to that associated with other aquatic rodents such as nutria and the North American beaver (see Chapter 8).

Marmots

Among the larger terrestrial rodents, the burrowing habits of members of the genus *Marmota* have been scrutinized both in North America and Eurasia. They are known as woodchucks (*Marmota monax*) in the agricultural areas of eastern North America, where their propensity for burrowing creates hazardous conditions, associated with cave-ins, for farm machinery and may lead to excessive aeration and damaging of plant roots (Swihart and Picone 1994). Marmots are steppe-dwelling animals in central Asia, whereas in western North America they are subalpine animals, establishing burrows on open, herb-covered talus slopes (Fig. 7.5) or grassy meadows in which rock outcrops and boulders are common (Svendsen 1976; Del Moral 1984). They also occur in the Alps, and were introduced into the Pyrenees at the end of the 1940s (Martinez Rica and Pardo Ara 1990).

The density of marmot burrows varies with environmental severity: In dry central Asia, burrow densities vary from about two per hectare in Kazakhstan for *Marmota bobak* (Rumyantsev 1989) to about seven per hectare for long-tailed or red marmots (*Marmota caudata*) in the eastern Pamirs (Tadzhiyev and Odinoshoyev 1987). Svendsen (1976) reported a density of more than eighty burrows per hectare for yellow-bellied marmots (*Marmota flaviventris*) in subalpine central Colorado.

The effects of marmots on soils, and therefore indirectly on surface runoff and erosion, have been described in several studies. A study in Connecticut suggests that no differences in slope, soil moisture, or soil pH exist between woodchuck burrow areas and nonburrow areas (Swihart and Picone 1994), but another in the eastern Pamirs of central Asia (Tadzhiyev and Odinoshoyev 1987, p. 22) illustrated that "the burrowing activity of marmots in high-altitude regions of the Tien Shen and Pamirs results in the redistribution of moisture, humus, carbonates, and salts in the soil profile and helps to change the genetic horizons and form the microrelief," and that soil texture was also altered. Del Moral (1984) also illustrated moisture redistribution caused by the grazing action of Olympic marmots (*Marmota olympus*) in the Olympic Mountains of Washington State. Svendsen (1976) noted that the soil at burrow sites in Colorado, similar throughout his study area, was a sandy loam containing rocks excavated from burrows. Individual excavated rocks ranged up to boulder size.

In spite of the fairly extensive knowledge of marmot habitat and burrowing activity, very few quantitative data exist on the actual amount of sediment removed from marmot burrows and subsequently eroded. Tadzhiyev and Odinoshoyev (1987) showed that the volume of soil removed from bur-

Figure 7.5. Marmot (arrow) excavating a den/burrow site at alpine tree line, Glacier National Park, Montana.

rows by red marmots in the eastern Pamirs ranged from 5 m^3 ha^{-1} in an "alpine cryogenic meadow" to 28 m^3 ha^{-1} in a high alpine desert. These data, in concert with information on the influence of marmots on soil characteristics, at least suggest that marmots can locally be major geomorphic agents across their geographic ranges in Eurasia and North America; however, given the widespread distribution and number of species of the genus, much more work remains to establish the importance of marmots in the context of alpine debris cascades in comparison to more quantified geomorphic processes such as soil creep, rockfall, snow avalanching, and debris flowage.

Prairie dogs

Prairie dogs are large (~1 kg as adults) herbivorous rodents that burrow and live in colonies known as "towns" (National Park Service 1981; Whicker and Detling 1988a). Their burrowing is so efficient, and was so geographically widespread prior to European colonization of North America, that it has been said that only the North American beaver (see Chapter 8) surpasses the engineering feats of prairie dogs (National Park Service 1981). They ranged over six hundred thousand square miles (McNulty 1970) in the early 1800s, and as late as 1919 covered more than 20% of the potential area of North American mid- and short-grass prairies, about forty million hectares (Whicker and Detling 1988a). Population estimates are subject to error even today (Powell et al. 1994), but "best guesses" place the pre-European settlement population of prairie dogs at between five billion and twenty-five billion individuals (McNulty 1970; National Park Service 1981)! Individual prairie dog towns covered enormous areas, the largest ever recorded being in Texas (of course!) at the beginning of the twentieth century. That town contained an estimated 400 million prairie dogs and covered an area 160 km wide by 400 km long. Like badger setts, prairie dog towns have been occupied in some cases for well over a hundred years (Carlson and White 1988; Burns, Flath, and Clark 1989)

Five species of prairie dog occur in North America, of which the black-tailed (*Cynomys ludovicianus*) and white-tailed (*C. leucurus*) prairie dogs are the most widespread (National Park Service 1981; MacDonald and Hygnstrom 1991). Blacktails inhabit the semiarid Great Plains region, and whitetails more commonly live in the higher elevations of mountain parks and foothills (National Park Service 1981).

Prairie dog burrows are approximately 10–13 cm in diameter, 1–5 m deep with two or three entrances, and 10–30 m long (Sheets, Linder, and Dahlgren 1971; Whicker and Detling 1988a; Cincotta 1989). Using those dimensions, Whicker and Detling (1988a) calculated that prairie dogs mix approximately 200–225 kg of soil per burrow system, much of which is deposited around burrow entrances in conspicuous spoil heaps. Densities of entrances vary, but estimates range from about 50 to 300 ha^{-1} (Whicker and Detling 1988a; Reading et al. 1989). Spoil mounds can reach 1 m high and 2.5 m in diameter (Cincotta 1989). The faunalturbation associated with the mounds increases soil texture variation as subsoil material is brought to the surface, and mound pH increases as calcareous parent material is deposited in the mound (Carlson and White 1988).

In spite of the astonishing statistics regarding prairie dogs in the preceding paragraphs, there has not been one single geomorphological study devoted to a more intensive examination of the quantitative effect of prairie dog burrowing. Consider also that, since the late 1800s, both the U.S. federal and state governments have actively exterminated prairie dogs at the behest of cattle-ranching interests, so that current populations, like those of the bison (Chapter 6) and beaver (Chapter 8), are at a fraction of their pre-European level. What effects have the removal of these pedoturbationally active, mound-building rodents had on bulk density of soil, infiltration capacity and surface runoff, and subsequent erosion? No one knows! Although one can speculate, the complex effects of prairie dog burrowing remain as of this time one of the most glaring omissions in a program of zoogeomorphological research. Whicker and Detling (1988a) emphasized the importance of the prairie dog as a keystone ecological species in Great Plains grasslands, and geomorphologists should attempt a determination if the same applies in our science.

Ground squirrels

The effects of burrowing by ground squirrels are visible on the landscape throughout the North American cordillera and the northern circumpolar region (Fig. 7.6). Most research has examined burrows of members of the genus *Spermophilus* (Smith and Gardner 1985; Laundré 1989, 1993; Young 1990; Yensen, Luscher, and Boyden 1991), although other genera have been studied (cf. Price 1971).

Because of the environmental harshness of the high latitudes and altitudes where ground squirrels such as *Spermophilus* live, most species hibernate for a full eight to nine months (Ritchie and Belovsky 1990). They therefore excavate both natal burrows and hibernacula (Young 1990; Laundré and Reynolds 1993). Most ground squirrels in the genus *Spermophilus* deposit excavated materials in spoil mounds (Fig. 7.6b), with pebble-, cobble-, and

Figure 7.6. (a) Columbian ground squirrel (*Spermophilus columbianus*) digging in a sub-alpine meadow. (b) Ground squirrel burrows and excavated soil material in the same sub-alpine meadow; Logan Pass area, Glacier National Park, Montana.

boulder-sized materials not uncommonly moved (Fig. 7.7) (Yensen et al. 1991).

From the perspective of assessing the geomorphic importance of ground squirrels, three studies stand out as noteworthy. In the first, Price (1971) ex-

Figure 7.7. Clastic material excavated by Columbian ground squirrel at a den site and deposited downslope of den entrance on a 30° slope, Glacier National Park, Montana.

amined the effects of the Arctic ground squirrel (*Citellus undulatus*) in the alpine tundra of the Ruby Mountains of Yukon Territory, Canada. Price noted a spatial relationship between the burrow entrances and spoil mounds of ground squirrels and the fronts of solifluction lobes on southern-facing slopes. He attributed this positioning to the more favorable microclimate of the warmer southern exposures, in concert with insulation from snow-drifting initiated by northwesterly winds onto southern aspects. The snow protected the squirrels during the harsh subarctic winter, yet snowmelt began sufficiently early to allow lobe fronts to thaw and provide a deeper active layer in which the squirrels can burrow.

Price itemized the direct and indirect effects of the ground squirrel burrowing in his study area. Directly, squirrels caused breaking away of vegetation; produced tunnels in the interior of solifluction lobes, which provided channels for meltwater that weakened the interior of the lobes; changed the soil bulk density; and actually transported material downslope in the burrowing process (cf. Figs. 7.6b, 7.7). Indirectly, the undermining of large rocks by burrows allowed the rocks to move more rapidly by rock creep or even tumbling downslope, and created conditions "favorable for the development of considerable hydrostatic and/or cryostatic pressure whereby material may be suddenly and forcibly moved downslope" (Price 1971, p. 102). Field measurements of the amount of material excavated and deposited on mounds revealed an impressive amount of 19.5 t ha^{-1} yr^{-1}. Price also noted that because the influence of the squirrels was limited to southern exposures, their actions may be significant in explaining the valley asymmetry frequently found in Arctic environments.

The second study of geomorphic note was carried out in the Mount Rae area of the Canadian Rockies of Alberta by Smith and Gardner (1985). They noted especially evident burrows and den entrances created by Columbian ground squirrels (*Spermophilus columbianus*) (see Fig. 7.6a) immediately above the upper forest limit (such as in Fig. 7.6b), particularly on south-facing slopes. They selected two isolated study sites, and for several years mapped and then weighed all of the debris contained within fresh burrow mounds produced each year. Their calculations of the volume of sediment added to each burrow mound each year ranged from 0.25 to 0.44 m^3 yr^{-1}. Those data led Smith and Gardner to calculate that 4–7 m of tunnel were added to each burrow system annually.

At the two burrow study sites, Smith and Gardner (1985) calculated the average rate of sediment transport to the ground surface at 1.35 and 1.12 t ha^{-1} yr^{-1}. This sediment was, in virtually all of the cases examined, pushed downslope at the burrow entrance into loose mounds subject to subsequent erosion. Smith and Gardner (1985, p. 209) noted that the calculated tonnage per hectare per year was "an order of magnitude less than that accomplished by solifluction and soil creep and several orders of magnitude less than that done by rockfalls, avalanches, solution, and streamflow suspended load." However, they also noted that their assessment of the work done by ground squirrels was based solely on the mound-building process; they did *not* consider the subsurface transfer of material by squirrels, nor the subsequent removal of mound material by other erosional agents. Given those limitations on their results, and given the spatial restriction of the ground squirrels to a narrow altitudinal belt in the alpine tundra and subalpine meadows, Smith and Gardner (pp. 209–10) conceded that "[w]ithin these restricted areas, however, the squirrels are probably the dominant geomorphic agents under existing environmental conditions." This conclusion is strikingly similar to that reached in my examination of the denning and food-digging activities of grizzly bears within the alpine tree-line ecotone (Butler 1992), just below the altitudinal zone dominated by ground squirrel activities.

The third study of particular interest examined the effects of ground squirrel (*Spermophilus townsendii* and *S. elegans*) burrows on water infiltration rates in the cool desert of Idaho (Laundré 1993). Laundré basically set out to actually test the oft-repeated assumption that animal burrowing significantly effects infiltration of water into burrow soils. Although his work was carried out from an ecological rather than a geomorphological paradigmatic framework, it clearly demonstrated the importance of burrowing. He compared the amount of water added to the soil profile from spring snowmelt recharge in areas subject to ground squirrel burrowing to that in nearby areas without burrows. His results illustrated that recharge amounts in burrow areas were significantly higher than in nonburrow areas, with an average of 21% more of the winter precipitation infiltrated into the soil near burrows. The amount of recharge was also related positively to burrow density. Furthermore, the burrows allowed vertical penetration of the water to deeper portions (>50 cm) of the soil profile than in nonburrow areas. Serendipitously, although Laundré's work was designed to examine the effects of spring recharge, four summer rainfall events >1.5 cm occurred at his study sites; for all of the rainfall events, significantly more moisture infiltrated into the top 10–30 cm of the soil in burrow areas than in nonburrow areas. Although Laundré did not assess the geomorphic impact of reduced surface runoff resulting from greater infiltration of burrow soils, the implications are that surface wash and fluvial erosion would be correspondingly reduced. In concert with the localized, but very important, effects of ground squirrel burrowing described by Price (1971) and Smith and Gardner (1985), it is clear that these small rodents have major geomorphic effects on the environment on which they live.

Kangaroo rats

Kangaroo rats are small, nocturnal burrowing rodents common in great numbers throughout the southwestern United States and northern Mexico (Dodge 1964; Moroka, Beck, and Pieper 1982). Most studies of kangaroo rat burrows have focused on those of the banner-tailed kangaroo rat (*Dipodomys spectabilis;* e.g., Best 1972; Moroka et al. 1982; Moorhead et al. 1988; Mun and Whitford 1990; Hawkins and Nicoletto 1992), although Best, Intress, and Shull (1988) compared the burrows of that species with those of a related subspecies and a separate species (*Dipodomys nelsoni*).

The burrowing activities of kangaroo rats leads to the production of distinctive mound topography (Best 1972) that may cover as much as 2% of some desert areas (Moroka et al. 1982). Mound densities typically range from about seven to sixteen per hectare (Moroka et al. 1982; Best et al. 1988). Typical mounds are 1–2 m in diameter and average about 15–30 cm in height (Lee 1986; Moorhead et al. 1988), although mounds of up to 60 cm high and 4–6 m in diameter are not uncommon (Mun and Whitford 1990; Hawkins and Nicoletto 1992). The average number of burrow entrances per mound ranges from between one and two up to over ten (Best et al. 1988). Lee (1986) used the method of multiple working hypotheses to illustrate that similar mounds in his study area, for which he did not actually observe kangaroo rat activity, were nonetheless attributable to their actions because of the incompatibility of mound morphology to alternative origins such as eolian deposition or fluvial and eolian erosion.

Kangaroo rat mounds have, as a result of the faunalturbational activities, better drainage and lower bulk density than surrounding desert soils (Lee 1986; Moorhead et al. 1988; Mun and Whitford 1990), although particle-size comparisons with surrounding soils reveal virtually no difference. Instead, the development of higher rates of infiltration and lower bulk density on the mounds is a result of more loosely aggregated soil structures (Moorhead et al. 1988). The net result is also a lower soil-water potential on mounds due to the high rates of infiltration in concert with high evaporation from plants finding favorable microhabitats on the mound soil (Mun and Whitford 1990). Interestingly, mounds with similar characteristics were reported by Cox (1987) from the Namib Desert of Africa; there, gerbils (*Gerbillus* spp.) rather than kangaroo rats were responsible, but the similarity in rodent morphology and life-style, as well as resultant landforms, is striking.

Mun and Whitford (1990) also pointed out that although most distur-bances in the desert of the U.S. Southwest and northern Mexico are short term and have few effects on soil characteristics, kangaroo rat mounds are long-term features acting to substantially alter the environment. Hawkins and Nicoletto (1992, p. 206) consider the kangaroo rat to be a "keystone species" in the semiarid grassland environment, noting that

[t]he banner-tailed kangaroo rat modifies its environment by building a structure that provides a relatively moist microhabitat which then is used by many other or-ganisms. The activity of this one species creates habitat patches that are physically and floristically different from the surrounding grassland and that become focal points of animal activity.

This conclusion certainly sounds similar to that associated with prairie dog activity elsewhere, in the semiarid grasslands of the Great Plains!

Moles

Moles are burrowing insectivores that otherwise live fossorial existences very similar to those of the gophers and mole-rats described in the next sec-tion. The European mole (*Talpa europaea* Linnaeus) and the American gar-den mole (*Scalopus aquaticus*) are probably the most common moles (God-frey and Crowcroft 1960), but attention from a geomorphic perspective has been restricted at this point to the European mole.

Moles are prolific burrowers, with some males estimated to burrow up to 20 m daily (Godfrey and Crowcroft 1960), although normal rates of digging are much lower. The burrowing activities of the mole are reflected on the surface as molehills, comprising the soil brought up from burrows and un-derground chambers. Molehills vary in size, but frequently range from 30 to 40 cm in diameter and 15–20 cm high (Jonca 1972). Molehills thrown up on slopes are especially susceptible to subsequent erosion by rainsplash and surface wash (Jonca 1972; Imeson 1976; Imeson and Kwaad 1976). The density of molehills varies considerably: Jonca (1972) noted a range of from less than twenty to several thousand per hectare.

Imeson (1976) examined the amounts of material raised to the surface by burrowing mammals in the Luxembourg Ardennes, as well as the rates at which it was raised. He found that the large majority was attributable to mole burrowing; the work of voles came in second. The rate at which mate-rial was raised to the surface by moles was measured at 1,940 m^3 km^{-2} yr^{-1}, which Imeson held to be underestimated, due to the overlooking of small mounds covered by plant litter. Although when expressed in terms of surface denudation these data revealed an extremely slow rate, averaging only 0.11 mm 1,000 yr^{-1}, two plots had significantly higher values of 3.4 mm and 30.6 mm per millennium. Imeson noted that these data placed the role of mole (and vole) burrowing in the same general range as Darwin's rates for earthworm-cast deposition on the surface (1,829–4,390 cm^3 km^{-2} yr^{-1}), but noted (Imeson and Kwaad 1976) that this role is less important than the moles' exposure of the material on the surface to subsequent splash erosion. Imeson (1976, p. 124) concluded by suggesting that the "burrowing activity of animals is one of the major slope processes occurring in humid deciduous forested areas today."

In contrast to the mild climate of Imeson's (1976) study area in Luxembourg, Jonca (1972) examined molehills in the harsh continental climate of the Karkonosze Mountains of Silesia, where snow cover is extensive in winter, with occasional midwinter thaws. Jonca noticed that molehills there frequently have a "tower" or "mushroom" shape, and that such molehills appear thorough the snow cover. Because such molehills are easily eroded by slopewash, Jonca was curious as to their mode of formation and their subsequent denudation. He explained their formation as follows:

[S]now falls at the beginning of winter before the ground freezes. At that time . . . moles are still burrowing in the soft ground and eject soil from their burrows on to the snow cover. In the study area, if the snow cover is slightly frozen on the surface, then waste slips a little down the slope, forming a mushroom of rock waste resting on a thin stalk. If a low temperature is maintained for a longer time, leading to gradual freezing of the soil, the molehill also freezes and is next covered by falling snow. It is not exposed until a temporary thaw occurs, or the spring thaw, but remains in a form similar to a mushroom. It is then more subject to degradation and denudation, as it has a lesser area of contact with the ground than the normal mound-shaped molehill [such as characterized Imeson's (1976) study area]. (1972, p. 411)

Jonca noted that during spring or temporary thaws, thermal degradation and gravitation shifting could remove 30–70% of a molehill's mass within two to three days of thaw. As a result of several recurrences of the process over the course of a winter, 70–80% of rock material forming molehills was denuded, amounting to an average of two tons per hectare (Jonca 1972). Despite the difference in molehill morphology, these data are strikingly similar to those reported by Imeson (1976).

Gophers and mole-rats

At least four distinct families of rodents are primarily composed of fossorial herbivores:

1. the Geomyidae (gophers, genera *Geomys, Thomomys,* and *Cratogeomys*) of North America (Hickman and Brown 1973; Teipner et al. 1983; Huntly and Inouye 1988; Lessa and Thaeler 1989),
2. the Ctenomyidae in South America,
3. the Spalacidae in Europe and the Near East (Heth 1991), and
4. the Bathyergidae in sub-Saharan Africa (Lovegrove and Painting 1987; Reichman and Jarvis 1989; Lovegrove 1991).

The latter two groups collectively comprise a group generically referred to as "mole-rats." Because little is known about the geomorphic effects of the Ctenomyidae in South America (but see Cox and Roig [1986] for an exception), this section focuses on the burrowing results of mole-rats and gophers.

Mole-rats are true subterranean rodents, spending most of their life in sealed underground burrows and coming to the surface only incidentally. They are prodigious burrowers, creating extensive tunnel systems and nest mounds. Excavated burrow systems in Israel averaged 40.3 m in length (Heth 1991). Nest mounds may be quite large, but last for only a few years before they are destroyed by surface wash and wind (Heth 1991). Nest-mound sizes reported in Israel ranged up to $160 \times 130 \times 40$ cm (Heth 1991). From three study sites in western Cape of Good Hope, South Africa, Reichman and Jarvis (1989) reported mound volumes of 1,470, 2,150, and 7,280 cm^3 ha^{-1}. Densities of mounds may reach in excess of twenty-five hundred per hectare (Reichman and Jarvis 1989). Indirectly, mole-rats affect the landscape by improving soil aeration and speeding nutrient cycling (Reichman and Jarvis 1989; Heth 1991), likely resulting in reduced bulk densities and increased infiltration capacities of the soil.

The efficiency with which mole-rats construct mound environments has led to the suggestion that they are responsible for the creation of Mima-like mounds in a diversity of landscapes in Africa (Cox and Gakahu 1983, 1984, 1985, 1987; Gakahu and Cox 1984; Cox, Lovegrove, and Siegfried 1987; Cox, Gakahu, and Waithaka 1989; Lovegrove 1991), although such an origin is by no means universally accepted (Ojany 1968; Martin 1988). The case of subterranean rodents and Mima-mound formation is examined later in this chapter (see "The question of Mima mounds").

Gophers of the family Geomyidae are distributed throughout the western two-thirds of the United States and Canada, northern Mexico, and in sandy-soil communities in the southeastern U.S. states of Alabama, Florida, and Georgia (Teipner et al. 1983). They produce vast underground tunnel and chamber systems, the design of which is apparently species dependent (Williams and Cameron 1990). In the course of their burrowing, they deposit tunnel casts and mounds on the ground surface, as well as subsurface stone zones ("stone lines") in coarse gravelly soils (Johnson 1989). Surface deposition of sediment reduces local vegetation cover, which acts in concert with complex effects on local soils (cf. Hansen and Morris 1968) to function ecologically as a disturbance that produces complex patch and gap-dynamic interactions with vegetation and other animals (Hirsch et al. 1984; Krylova and Deistfel'dt 1987; Gates and Tanner 1988; Huntly and Inouye 1988; Sparks and Andersen 1988; Peart 1989; Davis et al. 1991; Hobbs and Mooney 1991).

Northern pocket gophers (*Thomomys talpoides*) in particular have supplied interesting data on landscape ecological topics that indirectly influence geomorphic processes, because a population of these gophers survived the May 1980 eruption of the Mount St. Helens volcano in Washington State. Burrowing and mound construction by these eruption survivors modified the physical structure of the volcanic tephra (in some areas, gopher activity had covered up to 2% of the volcanic landscape within the first four months following the eruption; Andersen and MacMahon 1985a) and facilitated and influenced subsequent plant and animal succession in the volcano "blast zone" (Andersen and MacMahon 1985b, 1986). The prodigious nature of gopher digging was also reported by Thorne and Andersen (1990), who reported that a solitary pocket gopher excavated >110 m of tunnel and deposited 134 surface mounds during a 158-day period.

Specific geomorphic effects of gophers have been the subject of study for decades, but few precise quantitative data exist. Ellison (1946) reviewed the early literature, and measured the amount of soil deposited by gophers in the subalpine of the Wasatch Plateau of Utah. He reported an estimate (converted from English to metric units) of 11.0–14.5 t ha^{-1} yr^{-1}, with about 3.5% of the surface area covered by gopher spoils.

Burns (1979) and Thorn (1978, 1982) both examined the geomorphic effects of the northern pocket gopher (*Thomomys talpoides*) in the alpine tundra of Niwot Ridge in the Colorado Front Range. Burns reported that 91% of the burrow activity was spatially concentrated on the downwind slopes of large terraces where winter snowbanks form, positionally analogous to Price's (1971) observations concerning ground squirrels in the alpine tundra of the Yukon Territory. Burns also noted that the bulk of the gopher activity occurred during June–October, although sediment excavation and mound production occurred year-round. His measurements revealed a yearly sediment production of roughly a hundred thousand cubic centimeters per animal. His calculations showed a net erosion loss of excavation deposits of ~35% from each site, corresponding to an average surface lowering of

0.0037 cm yr^{-1}. By comparison, Burns noted that *nivation* processes (erosion effects associated with immobile, long-lived snow patches) in this study area accounted for surface-lowering values of 0.009 cm yr^{-1}, and "normal wind and water erosion on the alpine tundra" accounted for 0.0001 cm yr^{-1}.

Thorn's (1978, 1982) studies provided data similar to those cited by Burns, but his sediment-redistribution values of 3.9–5.8 t ha^{-1} were less than half those reported by Ellison (1946). One of Thorn's (1982) study sites did, however, provide an estimated absolute maximum disturbance rate of 27.7 t ha^{-1}. Certainly no major discrepancies exist between the values, given the vagaries of patterns of vegetation, slope, and snow distribution existing between the Wasatch Plateau and the Colorado Front Range. Thorn (1978) also illustrated that gopher tunnels and burrow entrances could, once abandoned, lead to burrow degradation and the development of micro-scale terracettes, "a microform which is extremely widespread in the alpine" (p. 184). Thorn's measurements of sediment directly removed by gophers provided values of 0.03–0.04 cm ha^{-1}, which when translated into Bubnoff units (Young and Saunders 1986), provides a value of 300 B (300 mm 1,000 yr^{-1}). That value was "an order of magnitude greater than the total rate in a nivation hollow and three orders of magnitude greater than the present estimates for the tundra in general" (p. 186). Thorn concluded by urging that geomorphologists should be "accepting gopher ecology as an integral geomorphic variable, rather than an aberration to be avoided in research site selection" (p. 186).

A similar conclusion was reached in a strikingly different climatic setting by Black and Montgomery (1991), who examined the geomorphic effects of burrowing activity of the gopher *Thomomys bottae* in the mild marine climate of Marin County, California, about 20 km north of San Francisco. They mapped all gopher mounds in a 4,200-m^2 drainage basin, and measured the volume of seventy-four mounds. Mound volumes averaged 1,100 cm^3, and 1,150 mounds were mapped. Sediment transport was measured to incorporate the number and volume of mounds as well as the distance of downslope displacement from mound centroids. Their calculations revealed a range (dependent on centroid method used) of average transport rates of 0.48–5.88 cm^3 cm^{-1} yr^{-1}; a reasonable overall estimate, they believed, was 0.91–2.17 cm^3 cm^{-1} yr^{-1}. These short-term rates did not account for the amount of sediment exported from the basin during the Holocene, which they attributed to variability of gopher activity over time, as well as to their inability to account for the amount of sediment transported either due to the collapse of old tunnels or through the system underground. With these caveats, Black and Montgomery concluded that gopher burrowing was the dominant mode of sediment transport currently active in their small study area.

It is obvious, then, that fossorial rodents can produce major displacements of sediment over short periods of time, and that these displacement activities result in the deposition of surface mounds of sediment (see, e.g., Fig. 7.7). Although these mounds are relatively small, several workers in the field of zoogeomorphology believe that such burrowing and mound-building actions can, given sufficient time, produce large-scale mounds (Fig. 7.8) variously called Mima mounds, pimple mounds, prairie mounds, hogwallow terrain, biscuit scabland, *heuweltjies,* and a host of other terms (Dalquest and Scheffer 1942; Arkley and Brown 1954; Scheffer 1958; Cain 1974; Cox 1989; Cox and Scheffer 1991; Lovegrove 1991). Mole-rats and gophers especially have been invoked as the burrowing agents responsible for the creation of Mima-like mounds, but the origin of such features remains controversial. The following concluding section of this chapter examines the nature and status of this controversy.

The question of Mima mounds

Mima mounds are earth mounds, approximately 20–30 m in diameter and 2 m high (Fig. 7.8), found throughout the western two-thirds of North America, west of the Mississippi River from southern British Columbia, Canada, to Sonora, Mexico (Mielke 1977; Cox 1984a,b; Cox, Gakahu, and Allen 1987; Cox and Hunt 1990a; Cox and Scheffer 1991). Seen from the air, they resemble hundreds to thousands of spheres partly buried in the ground (Cox and Scheffer 1991). They take their most common name, Mima mounds, from their type locality, Mima Prairie near Olympia, Washington (Dalquest and Scheffer 1942). Densities of Mima mounds range from between one and three per hectare in the Great Plains to over fifty per hectare in areas of California (Scheffer 1958; Ross et al. 1968; Mielke 1977; Cox 1984a, 1986, 1989, 1990b,c; Cox and Gakahu 1986; Cox, Gakahu, and Allen 1987). Mima-like mounds have also been reported from the highlands of Kenya in eastern Africa (Gakahu and Cox 1984; Cox and Gakahu 1985, 1987; Darlington 1985; Cox et al. 1989), southern Africa (Cox, Lovegrove, and Siegfried 1987; Lovegrove 1991), and Argentina (Cox and Roig 1986).

Mima mounds in North America rise only from shallow soils, either from those with a periodically waterlogged basement or from those underlain with a dense basement of bedrock, gravel beds, claypans, or cemented/calcrete hardpans (Cox and Scheffer 1991). Mounds are typically composed of unconsolidated fine soil varying in texture from loamy sands to clay

Figure 7.8. Mima-like mounds on a low-angle alluvial fan at the base of the Lemhi Range, Birch Creek Valley, Idaho.

loams. Small stones are more concentrated in mound soils than in inter-mound soils, but larger stones ranging in size up to cobbles and boulders are frequently exposed in the intermound zones, forming a variety of stone stripes and polygons, that is, sorted patterned ground (Cox and Allen 1987b; Cox 1990b; Cox and Hunt 1990b). Cross sections through mounds (including those shown in Fig. 7.8) reveal a lenticular (biconvex) outline (Dalquest and Scheffer 1942; Vitek 1978; Cox and Scheffer 1991).

Theories of origin of Mima mounds

As the terminology for Mima mounds reflects diversity, so do theories regarding their origin (Saucier 1991), which has been the subject of speculation and debate for nearly a century and a half. During this time, five serious scientific hypotheses of the origin of Mima-type mounds have emerged, of which four were discussed by Cox and Gakahu (1986), the fifth being a recent addition to the debate:

1. wind or water erosion (Cain 1974; Cox and Gakahu 1986 [and references therein]; Rostagno and del Valle 1988),
2. wind deposition (Cox and Gakahu 1986 [and references therein]),
3. periglacial freeze–thaw dynamics (Johnson and Billings 1962; Malde 1964; Vitek 1978),
4. soil translocation by fossorial rodents or other animals (Dalquest and Scheffer 1942; Scheffer 1958; Cox and Gakahu 1983, 1984, 1986; Lee 1986; Cox and Allen 1987a,b; Cox 1990b,c; Cox and Scheffer 1991), and
5. seismic activity (Berg 1990a,b, 1991; Cox 1990a; Saucier 1991).

Each theory is briefly outlined and discussed below. (Readers interested in early references to Mima mounds from the late 1800s and early 1900s are referred to the bibliographies of the publications listed above, most of which summarize those qualitative early descriptions.)

In the *wind- or water-erosion theory,* mounds are residual high points left by wind or water erosion. Mounds may have been protected by trees (Cain 1974; Cox and Gakahu 1986) or shrubs, or because of bedrock or substrate control on the development of the local drainage network. The erosion hypothesis, popular earlier in the century, has largely fallen into disfavor for reasons outline by Cox and Gakahu (1986).

The *wind-deposition hypothesis* (summarized by Lee [1986]) states that mounds are accumulations of fine-grained soil deposition by eolian action around and beneath clumps of vegetation. Cox and Gakahu (1986) point out

that mounds occur in many areas where winds are relatively light and variable, and would, if attributable to wind deposition, represent a past period
during which the carrying capacity of wind was accentuated. They also note
that this hypothesis cannot account for the frequently encountered small
rocks and pebbles within mound centers.

The *periglacial freeze–thaw hypothesis* attributes the mounds to soil-
convection and frost-sorting processes similar to those currently operative
and producing sorted and unsorted patterned ground in high alpine and arctic environments (Washburn 1980; Butler and Malanson 1989). Although
some Mima-like mounds are found in high-elevation environments where
current climatic conditions are marginally periglacial (Johnson and Billings
1962; Vitek 1978), most are found in climatic zones distinctly nonperi-
glacial in nature (e.g., Fig. 7.8), suggesting that the mounds formed under a
past, usually Pleistocene-epoch, colder climate (Malde 1964). Frost sorting
should concentrate small rocks in intermound areas with few on top of
mounds; however, Cox and Gakahu (1986) point out that many mounds
have relatively high small-rock contents compared to intermound areas, as
well as more on the actual mound top. For these reasons – as well as the unlikely occurrence of periglacial climates even under Pleistocene conditions
in many areas with mounds (such as southern California and Louisiana) –
the periglacial hypothesis is probably restricted in its applicability to fairly
narrow elevational or latitudinal zones that can be clearly shown as currently or formerly periglacial.

The *fossorial-rodent (Dalquest–Scheffer) hypothesis* states that Mima
mounds are the result of soil translocation by burrowing rodents. Dalquest
and Scheffer (1942) originally propounded the theory with specific reference to the work of gophers; Scheffer (1958) subsequently extended the idea
to fossorial rodents in general. The theory's greatest proponents have been
George Cox and his collaborators, who have specifically examined soil
translocation by gophers (Cox and Allen 1987a,b; Cox 1990b,c) and have
extended the hypothesis to non–North American mound landscapes attributed to burrowing by mole-rats (Cox and Gakahu 1983, 1984, 1987; Cox et al.
1989). Mielke (1977) also accepted the Dalquest–Scheffer hypothesis, and
provided a succinct summary of its mechanism. Cox and Gakahu (1986)
summarized the operation as follows: "mounds are formed by the centripe-
tal translocation of soil accompanying the outward tunneling of fossorial
geomyid pocket gophers from sites consistently representing their centers of
activity" (p. 487). The fossorial-rodent hypothesis explains the distribution
of small rocks in mound crests, as well as the larger rocks distributed in intermound areas (Cox and Allen 1987a,b; Cox and Hunt 1990a,b), although

some authors suggest that other, nonrodent, fossorial animals are responsible for mound development (Ross et al. 1968; Martin 1988).

The most vocal critic of the fossorial-rodent hypothesis has been Berg (1990a,b, 1991), who recently has put forth the fifth, seismic, hypothesis. His objections to the fossorial-rodent hypothesis include observations that gophers are widely distributed but mounds are not, and that gophers are missing from many mound sites, including those of the type area at Mima Prairie, Washington (Berg 1990b, 1991). He suggests that gophers and other fossorial animals occupy mounds because they offer the best habitat, not because they built the mounds. He offered the analogy that simply because gophers occupy mounds says no more about their role in mound development than to suggest that caves are constructed by the bats inhabiting them (Berg 1990b). He also questions why, if mounds are created by gophers, do we not find Mima-like mounds in various states of development, rather than in a uniformly large, well-developed size? Finally, Berg suggests that gophers are destructive agents degrading mound quality, rather than acting as agents of mound construction.

Berg's seismic hypothesis requires the same general conditions of a relatively planar, impenetrable substrate overlain by fine unconsolidated sediments. In addition, then, seismic activity is needed (Berg 1990a). He notes that the bulk of Mima-mound sites in North America and around the world coincide with seismically active locations, offering a better spatial coincidence than the distribution of mounds and gophers. Seismic activity leads to mound production as seismic waves cause sediment vibration where fine-grained material overlies a solid substrate. Wave vibrations in laboratory experiments carried out by Berg sort particle sizes also: The softer, fine-grained material forms mounds, and coarser-grained material works its way into the intermound spaces, thus alternatively accounting for the distribution of stone lines and nets surrounding some Mima mounds.

Conclusions

Burrowing by mammals is an extremely widespread geographic and geomorphic phenomenon, found in climatic zones ranging from arctic tundra to deserts and the tropics. Although quantitative data have been slow in coming, I have shown in this chapter that mammalian burrowing can displace and erode large amounts of sediment. Because of the spatially discrete distribution of mammal burrows (i.e., they are not ubiquitous on the landscape), the geomorphic influence by any one species or group of animals is spatially concentrated in localized settings. In those settings, however –

such as with ground squirrels above tree line or grizzly bears at tree line – mammalian burrowing may be *the* single most significant geomorphic process operating on the landscape.

The role of fossorial rodents is particularly widespread around the world, and such animals are capable of drastically altering a local environment. Are they, however, capable of producing widespread features such as Mima mounds? Berg's (1991) insightful comment regarding the lack of transitional forms of Mima mounds, from initial gopher or mole mounds through a range of intermediate sizes up to the "full-grown" Mima mound, does raise serious questions about the efficacy of the Dalquest–Scheffer hypothesis, at least as universally extended by Cox and associates. It is very likely that, as with periglacial frost sorting, some Mima-like mounds are indeed attributable to the action of fossorial rodents. However, as has frequently been shown in the history of geomorphology, the concept of a polygenetic continuum, where several diverse processes can ultimately result in landforms that look strikingly similar, may be fruitfully applied as well to Mima mounds. Fossorial rodents may make some of these mounds, but I seriously question whether all such landforms can be "laid at their feet." The next chapter, however, examines the geomorphic results of a rodent whose effects are widespread and unequivocal: the beaver.

8

The geomorphic influence of beavers

More than any other animal except humans, beavers geomorphically alter the landscape through their dam building and related activities. This chapter examines the geomorphic effects of beavers with special emphasis on the North American landscape, where beaver were historically widespread and numerous, and where they have recently reoccupied their historic range.

Beaver species and morphology

Two closely related species of beaver comprise the modern genus *Castor:* *Castor fiber,* the native European beaver, and *Castor canadensis,* the North American beaver (Williams 1988). Although a large body of literature exists on the ethology and morphology of *Castor fiber,* much more attention has been given to the geomorphic effects of beavers in North America. Studies of dam building and the geomorphic influences of beavers in Europe (cf. Wilsson 1971; Sinitsyn and Rusanov 1990; A. Meadows 1991; Zurowski 1992) also suggest that little if any difference exists in the constructional activities of the two species and the corresponding geomorphic results; accordingly, except where specifically noted, the following discussion focuses specifically on *Castor canadensis* and its geomorphic influence on the landscape.

The North American beaver (*C. canadensis*) is a large (adult mass > 15 kg), herbivorous, semiaquatic rodent with webbed rear feet, a flat and scaly-appearing tail, and long front incisors used as chisels for chewing through wood (Avery 1983; Novak 1987; Scheid 1987; Ryden 1989; Butler 1991a; Reynolds 1993). Trees are their primary food source: Those used for food and constructional activities vary by region and by availability of different tree species; where available, cottonwoods, birch, willows, poplars, and especially aspen are the preferred types (Warren 1927; Pinkowski 1983; Echternach and Rose 1987; Basey, Jenkins, and Busher 1988; Beier and Barrett 1989; Dieter and McCabe 1989b; Fryxell and Doucet 1991, 1993; Fryxell 1992), but beavers will utilize southern hardwoods and softwoods such as pine where necessary (Crawford, Hooper, and Harlow 1976; Shipes, Fendley, and Hill 1979; Edwards and Guynn 1984). By selectively foraging for specific species, beavers profoundly affect the composition of forests in the riparian environment (Barnes 1985; Barnes and Dibble 1988; Johnston and Naiman 1990b; Malanson 1993).

An adult beaver can chew through a tree with an 11-cm diameter in about 15 min (Scheid 1987). The actual size of trees brought down by beavers varies, but can be substantial (Fig. 8.1). Bailey and Bailey (1918) described a cottonwood tree >115 cm in diameter downed by beavers in Glacier National Park, Montana, and Dugmore (1914) described a >150-cm-diameter birch tree that was felled. Contrary to popular belief, beavers do not have unerring control on the direction of fall of trees, as evidenced by reports of beavers misjudging the direction and being killed by felled trees (Scotter and Scotter 1989).

Historical and modern distribution of the beaver

The beaver is well adapted to a variety of environments, ranging from the Arctic tree line of Canada and Alaska to the desert fringes of Mexico, and is found in all fifty United States and each Canadian province (Butler 1991a). Beavers are generally absent north of the Arctic tree line, in peninsular Florida, and in arid portions of the southwest without adequate surface water supplies (Duncan 1984; Medin and Torquemada 1988). Before the arrival of European settlers, the North American beaver population is estimated to have ranged from sixty to four hundred million (Novak 1987; Naiman, Johnston, and Kelley 1988; Butler 1991a) over a range of about fifteen million square kilometers. The decline of the beaver due to overharvesting and drainage of wetlands is well documented (Moore and Martin 1949; Naiman, Johnston, and Kelley 1988).

Since around the beginning of the twentieth century, stronger conservation laws and changing fashions have allowed beaver populations to rebound. Beavers now occupy virtually all of their former range, but at a density of only about 10% of the pre-European level; population estimates range from about six to twelve million beavers (Naiman, Johnston, and Kelley 1988). Estimates of current populations are difficult to verify, however, because of the difficulty in counting what is a largely nocturnal animal. Attempts at determining population levels in local areas based on the size of

Figure 8.1. A cottonwood tree chewed on and felled by beavers, Middle Fork of the Flat-head River, Montana. Lens-cap diameter is 49 mm. Wood is consumed by beavers and/or used for constructional purposes.

beaver food caches or measures of vegetation type and habitat suitability have met with mixed success (Slough and Sadleir 1977; Parsons and Brown 1978; Howard and Larson 1985; Broschart, Johnston, and Naiman 1989; Dubec, Krohn, and Owen 1990; Easter-Pilcher 1990; Osmundson and Buskirk 1993; Robel, Fox, and Kemp 1993). In some areas, population recovery has been documented using dendrochronological (Warren 1932; Wilde, Youngberg, and Hovind 1950; Lawrence 1952; Neff 1959; Bordage and Filion 1988) and/or aerial photographic and geographic information systems analyses (Dickinson 1971; Parsons and Brown 1978; Gotie and Jenks 1984; Bogucki et al. 1986; Remillard, Gruendling, and Bogucki 1987; Johnston and Naiman 1990c). The use of aerial photography and aerial surveys to determine census counts or colony densities is, however, best applied to northern or montane habitats where beaver ponds, lodges, and food caches are easily identified. Recently, Robel and Fox (1993) showed how much more accurate ground censusing was along midwestern plains-state rivers, where beavers occupy narrow riverine corridors that do not display characteristics easily identifiable from the air.

The recovery of the beaver from near extinction at the turn of the twentieth century has indeed been successful. This success has meant that their dam-building activities and attendant ponds are increasingly conspicuous (Naiman et al. 1994) and frequently at odds with human land-use activities (Hibbard 1958; Arner and DuBose 1978; Forbus and Allen 1981; Byford 1983; Johnson and Aldred 1984; Bullock and Arner 1985; Hedeen 1985; "The eager beaver" 1986; Shepherd 1986; Wigley and Garner 1986, 1987; Bown 1988; Ermer 1988; Willing and Sramek 1989; Gray 1990; Balch and Jones 1991; Butler 1991b; Glozier and Lee 1991; Smith and Peterson 1991; Dieter 1992; Fiddelke 1992; Bhat, Huffaker, and Lenhart 1993). The ponds are nevertheless viewed by some as highly beneficial from the perspective of wetlands habitat creation (Fig. 8.2) and habitat diversity (Hair et al. 1978; Rebertus 1986; Coleman and Dahm 1990; Medin and Clary 1990, 1991; Winkle, Hubert, and Rahel 1990; Leidholt-Bruner, Hibbs, and McComb 1992; Hammerson 1994).

Dam building and its geomorphic effects

In forested environments beavers build dams to create a pond environment in which to live. Beavers dig bank burrows and dens if population density is high, if surface streams are too large to dam, or if insufficient woody vegetation exists to provide for dam construction (Ffolliott, Clary, and Larson 1976; Bown 1988; Buech, Rugg, and Miller 1989; Dieter and McCabe

Figure 8.2. A riparian landscape near the headwaters of Nyack Creek, Glacier National Park, Montana, heavily utilized by beaver. Note the numerous canal pathways in the marshy environments at lower center and lower left.

1989a; Malanson and Butler 1990, 1991; Butler, Malanson, and Kupfer 1992; Butler and Malanson 1994). They have even been known to occupy caves if soft sediments for digging are not available (Gore and Baker 1989). At the extreme northern edge of their range, beavers in the Northwest Territories of Canada utilize scroll depressions adjacent to point bars in the Mackenzie River delta, where a cyclic pattern of creation of riparian habitat use, abandonment, and reoccupation has been documented (Gill 1972).

Although wood and brush gnawed down by the beaver and mud are the primary materials used in dam construction, beavers are also opportunistic; Warren (1905) and Mills (1913) described cases where beavers used logs brought down by snow avalanches as constructional material for dams, and it is not unusual for rocks and even human trash to be utilized (Warren 1927; Apple et al. 1985; Skinner et al. 1988). Pullen (1975) stated that mud, applied by the front paws of the beaver, is the most common material used in sealing the dams. The mud is concentrated along the proximal, upstream slope of the dam (Fig. 8.3).

Dam sizes and shapes vary (Figs. 8.4–8.6), but are typically arches that are concave upstream (Warren 1905; Dugmore 1914; Ives 1942). Typical dam sizes may be in the range of 15–70 m long and 1–2 m wide (Figs. 8.5, 8.6), but dams <1 m wide are not atypical (Townsend 1953; Fig. 8.4). Dam heights tend to be widely variable (McComb, Sedell, and Buchholz 1990; also cf. Figs. 8.3–8.6). Exceptionally large dams approach 200 m in length; the record is apparently a Montana beaver dam that was 2,140 ft (652 m) long (Ives 1942). Long dams may be more than 1 m wide along the dam crest (Butler 1991a). The nature of the constructional material may have some influence on dam size: Winkle et al. (1990) found that willows did not provide as sturdy a building material as did aspen, resulting in smaller dams with more ephemeral ponds where willow was the primary constructional material.

Active dams are maintained throughout the spring and summer (Townsend 1953) in cooler climates, and in the humid subtropical portion of the beaver's range, dam maintenance is a year-round activity (Pullen 1975). Active dams can last for decades or even centuries (Butler and Malanson 1994).

Damming of streams by beaver not only creates appropriate habitat for the animals, but in some cases can completely alter the drainage pattern of the local area. Dugmore (1914) provided maps illustrating the damming strategies whereby beavers had diverted a stream that had been flowing in a southeasterly direction. The dam placements effectively dried out the old streambed, created a pond that extended across the local drainage divide,

Figure 8.3. The upstream side of a beaver dam is an area of sediment concentration, emplaced by the beaver during construction, and by sedimentation into a pond. The pond here is shown at a low-water level in autumn.

(a)

(b)

Figure 8.4. (a) Beaver dams need not be large to be effective. The low dam on which the man stands impounds the lake visible in (b), on Red Eagle Creek delta, Glacier National Park, Montana.

Figure 8.5. Beaver dams are sturdy features, easily bearing the load of a full-grown person.

(a)

(b)

Figure 8.6. Beaver dams on Otatso Creek, Glacier National Park, Montana. Beavers are re-occupying this natural range: When visited in 1983 and 1987, no dams were present in the drainage. In 1988, the dam in (a) was discovered; in 1990, one dam downstream from this site was encountered. When visited in 1991, at least three additional dams, including the one shown in (b), had been constructed.

and with a series of four new dams regulated the pond outflow into two new streams flowing to the north and northwest.

Dam density per kilometer of stream varies widely throughout North America. Reports from coastal (Leidholt-Bruner et al. 1992) and semiarid eastern (McComb et al. 1990) Oregon cite low densities of 1.2 dams per kilometer and 1 dam per 7 km of stream, respectively. In subalpine valleys of Colorado, Ruedemann and Schoonmaker (1938) reported forty-six dams along a 5.75-mile stretch of stream, or about five dams per kilometer. Naiman, Johnston, and Kelley (1988) cited a figure of 2.5 dams per kilometer in the boreal regions of northern Minnesota and 10.6 dams per kilometer in southeastern Québec. Butler and Malanson (1994) encountered more than five dams per 200 m of stream in the Rocky Mountains of northwestern Montana, and Woo and Waddington (1990) counted an average of 14.3 dams per kilometer in subarctic wetlands along the western shore of James Bay in northern Ontario.

Woo and Waddington (1990) recognized four morphohydrological types of beaver dams: well-maintained dams, subdivided into

1. dams with stream-flow overtopping (*overflow dams*) and
2. dams with water funneling through gaps in the dam crest (*gap-flow dams*);

and weakened and decaying dams, subdivided into

3. *underflow dams* (water moves through the weakened bottom structure) and
4. *throughflow dams* (water seeps throughout the entire dam structure).

Woo and Waddington did not recognize any clear pattern of preferential occurrence of certain dam types along particular stream segments, but noted that most large ponds were impounded by overflow and gap-flow types.

The hydrological effects of beaver dams, following Woo and Waddington (1990) and Butler and Malanson (1994), include the following:

creation of ponds, diversion channels, and multiple-surface flow paths;

reduction of downstream discharge during dry periods below overflow and gap-flow dams;

alteration of discharge during high flow (overflow and gap-flow dams hold back water until the lowest point on dam crests are overtopped, whereas underflow dams dampen flow peaks and extend flow recession; throughflow dams are not effective in altering high flows); and

alteration of the overall water balance (evaporation is enhanced by pond creation, but is typically offset by reduction of water loss to runoff).

Beaver dams also expand the area of riparian habitat and flooded soils, in turn elevating the water table and recharging groundwater (Bergstrom 1985; Parker et al. 1985; Shephard 1986; Johnston and Naiman 1987; Naiman, Johnston, and Kelley 1988; Skinner et al. 1988; Hammerson 1994). In one particularly impressive case ("Beaver build large dam . . ." 1938), beavers dammed a 10-km-long glacial finger lake with a dam that was >40 m long and 0.5 m high, raising the water level of the entire lake slightly over 1 ft (~30 cm), causing a volumetric increase of roughly ninety-seven million cubic feet of water (Fig. 8.7). Recently, Naiman et al. (1994) described how beavers have converted 13% of the Kabetogama Peninsula of Minnesota to meadows and ponds over a 63-yr period.

In areas of appropriate lithology (i.e., carbonate terrain), beavers may un-knowingly act as agents of karstification. Cowell (1984) described two cases from Ontario where beaver ponds acted to concentrate water that subse-quently disappeared through karstic sinkholes in the base of the ponds. Dams and canals were then built to combat lowering water levels, but served only to concentrate more water and exacerbate the karstification. The bea-vers were ultimately unsuccessful in maintaining a pond environment be-cause of the karst drainage, and abandoned the areas.

The increased water table created by beaver impoundments has a second-ary indirect geomorphic and biogeographic influence alluded to by Naiman, Melillo, and Hobbie (1986). The elevated water table leads to saturation of the root zones and frequent tree death in a zone surrounding the pond (Fig. 8.8). Naiman et al. (1986) estimated that as much as 50–60% of the wood input into beaver ponds categorized as "wind-induced" resulted from such saturated conditions and mortality. No quantitative measures exist as to the amount of sediment introduced into a beaver pond by tree-tip uprooting, but could obviously (as in the case of Fig. 8.8) be substantial.

The beaver pond environment and associated geomorphic influences

The establishment of a pond by dam construction provides beavers with an environment in which they are relatively safe from predators. The pond also provides easy access to woody food and constructional material sources ringing the pond, and allows the beavers to transport woody material in the water easily. Beavers build their lodges in the pond, which in turn serves as a "defensive perimeter" around the lodges (Fig. 8.9). Lodges are construct-ed primarily of wood and mud (Mills 1913; Dugmore 1914; Warren 1927; Ryden 1989; and personal observations). An underwater outlet, essentially a burrow, leads from the pond into the interior of the lodge. The amount of

Figure 8.7. Beaver-dam construction can considerably elevate the water level of preexisting lakes. Here, the depth of Bowman Lake in northwestern Glacier National Park, Montana, a 10-km-long glacial finger lake, was increased ~30 cm by a dam at its mouth.

Figure 8.8. Elevated water tables caused by beaver-dam construction can lead to saturation of soils and rooting zones, resulting in uprooting by natural forces such as wind. This massive root-and-soil complex was recently uprooted along the fringe of the beaver pond shown in Figure 8.11a.

(a)

(b)

Figure 8.9. (a) A beaver lodge (center) in a beaver pond near Polebridge, Montana. Lodges are constructed from wood and mud, and may become sufficiently vegetated to achieve relative permanency on the landscape. (b) A beaver lodge and frozen pond during winter (18 January 1993). Note the canal at lower right leading to the underwater burrow opening into the lodge. The beaver dam forms the arcuate ridge at upper left of the pond.

sediment used in the construction of a typical beaver lodge has not been quantified anywhere in the literature; but suffice it to say that the combination of wood and mud used in a lodge produces a structure of great strength and potential longevity (Fig. 8.10), particularly when the soil is frozen (see Fig. 8.9).

Beaver canals are another significant geomorphic aspect of life in and around a beaver pond, as seen in Cowell's (1984) description of canal-assisted karstification. Canals vary in size from <1 m to well over 100 m in length, and are about 0.35–1 m or more in width (Butler 1991a; Butler and Malanson 1994). Although created primarily to transport logs to lodges (Berry 1923; Warren 1927; Townsend 1953) (Fig. 8.11a), they are used also to divert water to maintain pond depth (Cowell 1984; Rebertus 1986), and to provide access to bank burrows (Fig. 8.11b). Some cases are known where beavers actually banked both sides of a canal so that conspicuous levees were constructed (Berry 1923); these allow the canal to fill to a distinctly deeper level.

The amount of sediment excavated by beavers for canal construction has not been quantified, but could be calculated in individual cases. For example, Cowell (1984) described one beaver canal that was about 1 m deep and 160 m long. If a conservative 50-cm width is assumed, based on a photograph in Cowell's paper, a value of 80 m^3 of excavation can be calculated for this one canal. Much more work needs to be done on quantifying the sediment-moving capabilities of beavers in building and maintaining canals.

Around the fringes of a pond, beavers may construct several bank burrows (Fig. 8.12), in some cases associated with canals leading from the main portion of the pond (Fig. 8.11b). Bank burrows are especially important in environments where conditions are not conducive to dam and pond construction, and where temperature extremes would otherwise preclude exploitation of a physiologically unsuitable environment (Ffolliott et al. 1976; Buech et al. 1989). In such environments, bank burrows are excavated into streambanks (Figs. 8.13, 8.14). Beaver "runs," or heavily trampled trails, frequently extend from bank burrows inland to feeding areas (Fig. 8.15).

Few data exist on the density or dimensions of burrows per pond or stretch of stream channel. Ffolliott et al. (1976) described a beaver pond in north-central Arizona that had twenty-seven bank burrows around its periphery. Butler and Malanson (1994) reported that bank burrows may extend several meters into stream or pond edges, disrupting tree root systems and potentially assisting the development of tree-tip mounds such as described above. One of the burrows illustrated in Figure 8.12 extended a minimum of 175 cm into the pond bank, was 25 cm high and about 70 cm

Figure 8.10. Similar to beaver dams, beaver lodges are sturdy constructional features comprised primarily of wood, sticks, and mud. The beaver lodge in this view had been abandoned for at least two years prior to its ascent by the person shown.

(a)

(b)

Figure 8.11. (a) Beaver (*Castor canadensis*) pond with extensive canal network; unnamed pond be- tween Red Eagle and Divide Creek drainages, Glacier National Park, Montana. (b) Beaver canal in lower right leads to bank burrow directly beneath the person; Otatso Creek drainage, Glacier National Park, Montana.

Figure 8.12. Draining of a beaver pond reveals the burrow entrance to the lodge in the pond center, as well as entrances to several bank burrows (arrows) along the pond's edge.

Figure 8.13. The arrow points to a beaver bank burrow under the overhanging river bank; Middle Fork of the Flathead River, Montana.

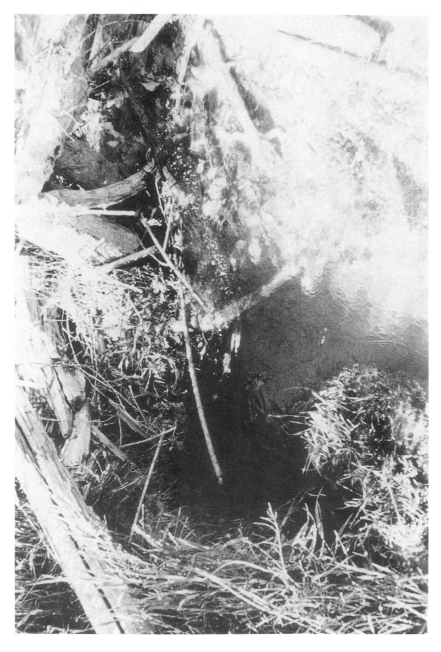

Figure 8.14. A closeup of the beaver bank burrow shown in Figure 8.13. Note the gnawed wood fragments in the water.

Figure 8.15. Beaver "runs," or heavily trampled trails, frequently extend from a feeding area to pond or stream edges.

wide, and therefore displaced into the pond over 0.3 m^3 of sediment (Fig. 8.16).

Conditions necessary for the construction of bank burrows have been described by Bown (1988) and Dieter and McCabe (1989b), and were recently summarized by Butler and Malanson (1994). Of primary importance is a substrate into which the beavers can easily burrow, that is, fine-grained alluvium. Beaver burrows are typically absent where pond edges or streamside alluvium is gravelly or rocky. Bank slopes may vary, but another factor of primary importance is sufficient water depth to mask the underwater entrances that typify bank burrows (Butler and Malanson 1994).

Sedimentation and sedimentation rates in beaver ponds

From the earliest geomorphic literature on beaver ponds, it has been suggested correctly that their ultimate fate is infilling with sediment over the course of decades to centuries (Warren 1905; Ruedemann and Schoonmaker 1938; Ives 1942). This infilling occurs because beaver dams reduce the ability of a stream to transport sediment by lowering the effective slope of the stream channel (Apple et al. 1985), with the beneficial side effect of the release of significantly cleaner water below the dam (Taylor 1985; Skinner et al. 1988). A series of beaver dams and ponds acts as a sequence of settling pools, stepping down flow velocity and thereby reducing erosion potential, instead inducing sediment deposition (Duncan 1984; Apple et al. 1985; Bergstrom 1985). The term "beaver meadow" has been applied to organic-rich meadows presumably formed in this fashion; in the southwestern United States, the term "vega" refers to the same feature (Dalquest, Stangl, and Kocurko 1990). Ruedemann and Schoonmaker (1938, p. 525) went so far as to suggest that "the fine silt gathered in the beaver ponds has produced the rich farm land in the valleys of the wooded areas of the northern half of North America."

Ives (1942) provided one of the first in-depth descriptions of beaver-pond sediments. He described them as "the normal gravels, sands and silts of any drainage area, with the usual admixture of organic debris. Very shortly after deposition, however, the sediments are firmly bound together by the roots of grasses and swampland brush, forming a matted zone of material which is surprisingly resistant to further erosion. Accumulations of humic acid in the subsurface portions of this pond fill inhibit decay, so that the filled beaver pond becomes a lasting feature of the valley floor" (pp. 196–7). The sediments described by Ives were deposited in a typically deltaic fashion with bottom, foreset, and topset beds.

Figure 8.16. The person stands in a collapsed beaver bank burrow located along the shore of the drained pond shown in Figure 8.12.

Dalquest et al. (1990) suggested that the sediments deposited in the deeper waters of beaver ponds tend to be stratified, dark in color, and rich in fossil shells, whereas vega sediments are largely deposited above the water table, initiating swift decomposition of organic matter and resulting in pale and unstratified sediments. Rains (1987) corroborated this description of buried pond sediments, and described the actual sediments comprising a buried mid-Holocene beaver dam: Gravel and finer-grained alluvium formed a matrix for the wood of the dam, which rested on a sharp erosional contact with underlying bedrock (Rains 1987, p. 272). Buried and preserved beaver dams were also briefly described by Ives (1942).

Sediment accumulation in beaver ponds, and in streams above ponds, provides anoxic conditions suitable for significant genesis of methane. Although not specifically geomorphic in nature, many studies of methanogenesis, nitrogen cycling, and elemental concentration in the sediments of beaver ponds also provide useful geomorphic information on the effects of damming, ponding, and sediment deposition (Naiman and Melillo 1984; Francis, Naiman, and Melillo 1985; Ford and Naiman 1988; Yavitt, Lang, and Sexstone 1990; Balch and Jones 1991; Naiman, Manning, and Johnston 1991; Yavitt et al. 1992; Bubier, Moore, and Roulet 1993; Naiman et al. 1994).

Amounts of sediment deposited in beaver ponds

Although the general formation of meadows through beaver pond infilling has been accepted in the literature, very little is known about the *amount* of sediment retained in beaver ponds, or the *rate* of accumulation and infilling. Coleman and Dahm (1990, p. 299) briefly described sedimentation in a drainage basin in the Zuni Mountains of New Mexico, where one high-flow event deposited "several cubic meters of sediment and organic matter behind the uppermost dam in Castor Creek." Naiman and colleagues (Naiman et al. 1986; Naiman, Johnston, and Kelley 1988) measured the amount of sediment retained behind several dams in Québec, Canada. They determined that there was no relationship between the amount of sediment retained and the size of the beaver dam ($r^2 = 0.03$, $n = 18$, $p > .05$), but that a significant relationship existed between the amount of sediment retained and the surface area of the pond (sediment volume = 47.3 + 0.39 $*$ [surface area]; $r^2 = 0.85$, $p < .01$, where sediment volume is in cubic meters and surface area is in square meters; Naiman et al. 1986, p. 1258). Their sample included pond sizes of ~100–14,650 m^2 and sediment volumes of ~35–6,500 m^3. A small dam of 4–18 m^3 of wood could entrap 2,000–6,500 m^3 if properly positioned (Naiman et al. 1986).

Naiman et al. (1986) did not describe rates of sedimentation, but they did examine the question of total sediment accumulation in beaver ponds at the landscape scale in their study area (the Matamek River watershed in Québec). By knowing the overall channel lengths of less-than-fourth-order streams, and by assuming ten beaver dams per square kilometer at 1,000 m^3 of sediment retained apiece, they extrapolated that beavers were directly responsible for the retention of 3.2 × 10^6 m^3 of sediment in small-order streams in the 673-km^2 study area – enough to cover the bottom of every surface stream in the drainage with an additional 42 cm of sediment!

Rates of sedimentation in beaver ponds

Little geomorphic attention has been paid to the rates of sedimentation in beaver ponds. Warren (1927) described buried archeological remains from a peat bog that formed from a beaver meadow. He calculated a rate of growth of the bog at about "a foot a century" (Warren 1927, p. 50), but did not provide an explanation for assuming such a rate.

Ruedemann and Schoonmaker (1938, p. 525) described beaver ponds that were abandoned by their occupants in 1903 or 1904. By 1912, meadows were "pretty well formed," and by 1921 the ground was solid and evidence of the dam(s) nearly gone. They concluded that a beaver meadow can, in an abandoned pond where dam effectiveness remains strong, form in about fifteen years. They did not cite evidence of the depth of sediment deposited, however, nor did they calculate an average rate of sedimentation.

Ives (1942), examining increases in water-level elevation of streams caused by beaver damming, noted that pond infilling occurs at about the same rate as the elevation of water levels but with a lag time of several years. He concluded that "valley floor elevation" – presumably synonymous with pond-infilling – proceeded at a rate of about one-quarter inch per year, but also noted that local residents in his study area in Colorado favored an annual rate of one inch (2.54 cm).

Recently, in collaboration with Dr. George P. Malanson of the University of Iowa, I have begun an examination of the average rate of sedimentation in a beaver pond of known age in southeastern Glacier National Park, Montana (Fig. 8.17). When we first visited this pond in August 1991 (Fig. 8.18), it had been drained by the National Park Service just that summer because of potential inundation of an adjacent park access road. The pond is visible on a late-summer 1990 Landsat Thematic Mapper image of the park, but *not* on a late-summer image from 1988; therefore, it had been no more than three years old when drained in 1991, and possibly only one year old.

Figure 8.17. The arrow points to a drained beaver pond site (see Fig. 8.12) on a delta protruding into Lower Two Medicine Lake, Glacier National Park, Montana. Beavers frequently utilize, and in turn enhance, deltaic environments.

Figure 8.18. Closeup of a recently drained beaver pond and dam that had threatened to inundate the road shown in Figure 8.17. Note the relatively uniform stick length and thickness used in dam construction.

Figure 8.19. The surface of the drained beaver pond when visited in October 1993, at least two years after pond drainage and lodge abandonment.

The surface area of the drained pond in October 1993 (Fig. 8.19) measured 1,710 m^2. At several locations across the surface of the drained pond, we recorded depth of penetration through fine sediments and took a representative sediment core (Fig. 8.20). The average depth of penetration, and the depth of the core, was 20 cm. This pond had, therefore, accumulated approximately 342 m^3 of sediment in 1–3 yr. Assuming a life span of three years provides an annual sedimentation depth rate of not quite 7 cm yr^{-1}, corresponding to an annual volume of 114 m^3. A two-year pond would have seen 10 cm yr^{-1} depth accumulation and a volume of 171 m^3 yr^{-1}.

The sediments exposed on the floor of the drained pond (Fig. 8.21) were of fine grained and relatively uniform particle size. They overlay a buried surface of rounded rocks and gravel that we have tentatively interpreted as fluvial in origin. Future work by our research team will investigate the type, depth, and rate of sediment accumulation in a variety of beaver-pond sites to examine spatial variations in those factors (Fig. 8.22).

Catastrophic dam failure and its effects

When high waters threaten the integrity of a beaver dam, beavers respond by attempting to relieve pressure on the dam by creating a small channel opening in its crest (Mills 1913; Dugmore 1914; Bailey 1922, 1927; Warren 1927; Wilsson 1971; Allred 1986). Overflow and throughflow may suffice to relieve pressure and allow the dam to withstand the high-water event (Woo and Waddington 1990), but beaver dams occasionally do fail catastrophically, leading to profound effects on downstream geomorphology and biota. Beaver dams break when stream discharge exceeds a critical strength threshold (Parker et al. 1985). Natural agents that may lead to dam failure include excessive precipitation over a short period of time (Dugmore 1914; Warren 1927, 1932; Rutherford 1953; Townsend 1953; Butler 1989, 1991a,b; Kondolf et al. 1991; Schipke and Butler 1991; Stock and Schlosser 1991; Marston 1994), rapid snowmelt (Warren 1927), otters burrowing through the dam (Reid, Herrero, and Code 1988), and collapse of upstream beaver dams, leading to a "domino effect" on downstream dams (Marston 1994). Beavers also periodically breach their own dams (Mills 1913; Warren 1927) in order to drain the water from their ponds. Warren (1927) stated that he did not know why beavers do this, although he cited Mills's (1913) speculation that it is done for sanitary reasons, that is, "to permit the sun and air to purify the bottom of the pond and interior of the lodge, if one was present" (Warren 1927, p. 46).

The amount of precipitation necessary to produce catastrophic dam fail-

Figure 8.20. Sampling sediment depth with a crude coring device in the drained beaver pond shown in Figure 8.19.

Figure 8.21. A soil pit in the drained beaver pond (Fig. 8.19) revealed ~20 cm of fine-grained sediment overlying coarser fluvially deposited alluvium. The contact point occurs at the distinct break in tone near the '7' (ruler shown is in inches).

Figure 8.22. Sediment sampling from additional beaver ponds of known age, such as shown here, will allow reconstruction of rates of sediment accumulation.

ure and flash flooding will, of course, vary with local topographic and vegetative conditions, and few studies have examined the question of the minimum precipitation necessary to exceed a dam's critical strength (Parker et
al. 1985). Kondolf et al. (1991) noted that dam washout along streams in
the eastern Sierra Nevada of California was associated with a year in which
runoff was 154% of normal. Stock and Schlosser (1991) reported that a beaver dam failed catastrophically after 156 mm of precipitation fell in a five-
day period in a forested catchment in Minnesota. Butler (1989) examined
five case studies of catastrophic dam failure on granitic terrain in the Piedmont of Georgia and South Carolina, concluding that dam failure was likely
if a threshold of 75–80 mm of precipitation was exceeded in a 24-hr period.

Although stream-discharge values associated with floods that washed out
beaver dams have been reported (cf. Rutherford 1953), discharges associated with catastrophic failure of dams and rapid beaver-pond drainage have
rarely been directly measured or calculated. Stock and Schlosser (1991) described a flood from catastrophic beaver-pond drainage that caused a stream
with a midsummer depth of 11–18 cm and a width of 3 m to increase in size
to 102 cm × 13.5 m; unfortunately, no velocity values were provided from
which discharge could be calculated. Butler (1989) calculated the discharge
for the five cases of dam failure and pond drainage in Georgia and South
Carolina, using pond volume and Costa's (1985) empirical equation for
landslide–dam failure in four cases, and Leopold and Maddock's (1953)
continuity equation ($Q = WDV$) in the fifth. In the first four cases, calculated
discharges ranged from 28 to 62 m^3 sec^{-1}. In the fifth case, the volume of
the failed pond was not known. The width and depth of the stream below
the pond at peakflow were determined from the distribution of flood debris
along the lateral margins of the creek (Figs. 8.23a,b), and velocity was determined from examination of the size of clasts moved and from the use of
sediment competence curves. In 1988 I measured a normal late-winter discharge for this stream at 0.28 m^3 sec^{-1} (Butler 1989; Fig. 8.23a). Calculations of peak discharge resulting from the beaver-dam outburst flood ranged
from 72 m^3 sec^{-1} to a probably more realistic 325 m^3 sec^{-1}. This flood
transported quarried granite clasts 1 m in diameter, swept a light passenger
truck downstream ~100 m, killed four people, and deposited a survivor 4 m
up in a tree along the channel margins; complete details are provided in
Butler (1989) and Schipke and Butler (1991).

In addition to the transport of unusually large clasts >1 m in diameter,
described by Butler (1989), beaver-dam outburst floods have profound impacts on stream morphology and biota. Kondolf et al. (1991) described the
results of a washed-out beaver dam on Tinemaha Creek in the Sierra Neva-

Figure 8.23. (a) Beaver dams are washed out during spring runoff in some locations in mountainous topography. They may also fail catastrophically as a result of torrential rains. The man stands ankle-deep in a stream that flooded as a result of a beaver-pond outburst. The resulting flood depth reached to the dashed line in the picture, above the truck. (b) The debris shown here resulted from the same beaver-dam failure as that shown in (a). The flood killed four people and reached the level of the dashed line above the head of the woman near upper left.

(a)

(b)

da of California; there, the washout lowered local base level and induced incision of ~0.5–0.6 m depth across the entire channel width. Gravels deposited against the upstream face of the beaver dam were swept away, "leaving a bed of cobbles and boulders and banks of exposed willow roots" (Kondolf et al. 1991). Smaller-grained clasts are also washed out and transported downstream when beaver dams fail, and silt deposition in the channel below the dam may be sufficient to smother benthic organisms and fish eggs (Rupp 1955). Stock and Schlosser (1991) illustrated a >90% decrease in benthic invertebrate densities below a beaver dam that failed catastrophically, and the postflood recovery was noticeably slow. However, the removal of the dam as a barrier to fish migration resulted in an increased population of fish for about two to four days following the outburst flood.

Conclusions

The removal of a beaver dam by catastrophic outburst flooding may be viewed as a discrete but unpredictable disturbance event (*sensu* Resh et al. 1988) in the context of local geomorphology and landscape ecology. As dams become more frequent, particularly in mountain valleys (Butler et al. 1991; Butler and Schipke 1992; Butler and Malanson 1994), the threat of dam outburst floods must increase. The dams themselves serve as distinct ecotones (Stock and Schlosser 1991) or new habitats (Clifford, Wiley, and Casey 1993; Medwecka-Kornas' and Hawro 1993), just as do the beaver ponds and the associated stream channels. As such, the dams, ponds, and stream environments (*sensu* Johnston and Naiman 1987, 1990a,b; Remillard et al. 1987; Naiman, DeCamps, et al. 1988; Naiman, Johnston, and Kelley 1988; Pringle et al. 1988; Smith et al. 1991; Johnston et al. 1993; Mitchell and Niering 1993; Stanford and Ward 1993), should *each* be considered as patch bodies in a landscape ecological paradigmatic framework, with the beavers themselves serving as an ecological keystone species (Naiman et al. 1986; Nummi 1989). In a geomorphic context, an enormous amount of work remains to be done before the effects of beavers on the landscape can be quantified. Measurements of beaver-landform morphometry should be undertaken, in concert with measurements of sedimentation rates in beaver ponds. It is an accepted paradigm that beaver ponds trap sediment and reduce stream erosion; but no one knows how much of the sediment introduced into such a pond, and ultimately compacted into beaver meadow sediments, is actually introduced *not* by the incoming stream(s) but by the digging efforts of the beavers themselves. Such questions await the efforts of future zoogeomorphologists.

9
Concluding remarks

In Chapter 1, I posed the question as to whether the geomorphic role and effects of animals are significant and fundamental, or merely interesting but minor curiosities. Thereafter, I examined the geomorphic role of invertebrates, ectothermic vertebrates, birds, and mammals. In *each and every case,* despite human interference in the natural life cycles and distributions of animals around the globe, the answer has been resoundingly in favor of significance, whether it be as a result of the work of termites in the tropics, or of whales and walrus on the Arctic seafloor.

Although an individual species or genus may not have geographically widespread geomorphic influence, it may have profound effects locally. Examples include the setts of European badgers, mounds constructed by crayfish, salmonid redds, or food and den excavations by grizzly bears at alpine tree line. One must keep in mind, however, how *many* species of animals produce geomorphic effects on at least a local scale. Individual studies that denigrate the geomorphic contributions of animals in comparison to the work of running water, wind, or mass movements invariably acknowledge the *local* importance of one or a few species of animals as geomorphic agents; but what those studies fail to recognize is the quantity, significance, and geographical ubiquity of geomorphic accomplishments by animals *collectively as a group.* Although the amount of sediment displaced annually by rabbits or badgers, for example, may be impressive but spatially restricted, what about all the other animals also working that same area, as well as areas beyond the influence of the rabbits or badgers? Surely earthworms and ants are also operative in many such locales, as are fossorial rodents, and perhaps birds collect mud for nests there. The geomorphic effects of animals must be considered *collectively* in order truly to be appreciated.

Consider also the arrogant perspective of humanity in its assessment of animals as geomorphic agents. We have completely exterminated a number of animal species and severely limited the number and geographic extent of others. How can we say that the American bison is not an important agent of geomorphology when we are assessing it from the context of having ourselves *removed* that agent in the first place? The same could be said for the large grazing mammals of the African savannas. To belittle the contributions of burrowing fish, amphibians, and reptiles is to deny our own roles in limiting those roles via destruction of their habitat and prey. Consider also the example of the beaver in North America. Before European colonization, beavers occupied virtually every corner of the continent except the harshest deserts. Within three hundred years, the beaver was nearly extinct, removed as a major geomorphic agent from across the continent. Today, in the late twentieth century, beavers have rebounded to perhaps one-tenth their pre-contact population. We cannot truly assess their significance as geomorphic agents unless we take that history into account.

Because many studies of the geomorphic role of animals are conducted in nature preserves, we delude ourselves into thinking that we are accurately assessing those animals' environments and processes. Certainly something can be learned about processes and rates in such controlled environments, just as we can learn about the meandering of rivers through the use of stream tables. However, truly to understand the river we must consider the roles of a suite of variables: upstream drainage-basin characteristics, substrate and sediment size, surface vegetation, and so on. We therefore study and attempt to measure geomorphic processes in the field, even while acknowledging that anthropogenic influences may alter the operation of the process. By comparison, however, when studying animals in the field we have in many cases exterminated the very agents whose geomorphic contributions we seek to understand.

So what of the future of zoogeomorphological studies, evoked at several points throughout this book? I have offered suggestions as to realms of research that could prove fruitful, or where more detailed work remains to be done. Perhaps the greatest part that geomorphologists could play in understanding the role of animals is to help *preserve* those animals. Only then will we ever have a true understanding of their fascinating, ubiquitous, and fundamental role in the science of geomorphology.

References

Aalders, I. H., Augustinus, P. G. E. F., and Nobbe, J. M. 1989. The contribution of ants to soil erosion: A reconnaissance survey. *Catena* **16**: 449–59.

Abaturov, B. D. 1972. The role of burrowing animals in the transport of mineral substances in the soil. *Pedobiologia* **12**: 261–6.

Able, K. W., Grimes, C. B., Cooper, R. A., and Uzmann, J. R. 1982. Burrow construction and behavior of tilefish, *Lopholatilus chamaeleonticeps,* in Hudson Submarine Canyon. *Environmental Biology of Fishes* **7**: 199–205.

Ahnert, F., ed., 1989. *Landforms and Landform Evolution in West Germany. Catena (Supplement)* **15**: 1–347.

Akpan, E. B. 1990. Bioerosion of oyster shells in brackish modern mangrove swamps, Nigeria. *Ichnos* **1**: 125–32.

Alkon, P. U., and Olsvig-Whittaker, L. 1989. Crested porcupine digs in the Negev desert highlands: Patterns of density, size, and longevity. *Journal of Arid Environments* **17**: 83–95.

Allred, M. 1986. *Beaver Behavior – Architect of Fame and Bane.* Naturegraph Publishers, Happy Camp, Calif.

Andersen, D. C. 1987. Below-ground herbivory in natural communities: A review emphasizing fossorial animals. *Quarterly Review of Biology* **62**: 261–86.

Andersen, D. C., and MacMahon, J. A. 1985a. The effects of catastrophic ecosystem disturbance: The residual mammals at Mount St. Helens. *Journal of Mammalogy* **66**: 581–9.

Andersen, D. C., and MacMahon, J. A. 1985b. Plant succession following the Mount St. Helens volcanic eruption: Facilitation by a burrowing rodent, *Thomomys talpoides. American Midland Naturalist* **114**: 62–9.

Andersen, D. C., and MacMahon, J. A. 1986. An assessment of ground-nest depredation in a catastrophically disturbed region, Mount St. Helens, Washington. *Auk* **103**: 622–6.

Andersen, F. Ø., and Kristensen, E. 1991. Effects of burrowing macrofauna on organic matter decomposition in coastal marine sediments. *Symposium of the Zoological Society of London* **63**: 69–88.

Andersson, S. 1989. Tool use by the fan-tailed raven (*Corvus rhipidurus*). *Condor* **91**: 999.

Apple, L. L., Smith, B. H., Dunder, J. D., and Baker, B. W. 1985. The use of beavers for riparian/aquatic habitat restoration of cold desert, gully-cut stream systems in southwestern Wyoming. In G. Pilleri, ed., *Investigations on Beavers,* vol. IV. Brain Anatomy Institute, Berne, pp. 123–30.

Arkley, R. J., and Brown, H. E. 1954. The origin of Mima mound (hogwallow) microrelief in the far western states. *Soil Science Society of America Proceedings* **18**: 195–9.

Arner, D. H., and DuBose, J. S. 1978. Increase in beaver impounded water in Mississippi over a ten year period. *Proceedings of the Annual Conference Southeastern Association of Fish and Wildlife Agencies* **32**: 150–3.

Arshad, M. A. 1981. Physical and chemical properties of termite mounds of two species of *Macrotermes* (Isoptera, Termitidae) and the surrounding soils of the semiarid savanna of Kenya. *Soil Science* **132**: 161–74.

Atkinson, R. J. A., and Taylor, A. C. 1991. Burrows and burrowing behaviour of fish. *Symposium of the Zoological Society of London* **63**: 133–55.

Avery, E. L. 1983. *A Bibliography of Beaver, Trout, Wildlife, and Forest Relationships.* Wisconsin Department of Natural Resources Technical Bulletin No. 137, Madison.

Avery, W. E., and Hawkinson, C. 1992. Gray whale feeding in a northern California estuary. *Northwest Science* **66**: 199–203.

Ayeni, J. S. O. 1977. Waterholes in Tsavo National Park, Kenya. *Journal of Applied Ecology* **14**: 369–78.

Babcock, E. A. 1976. Bison trails and their geological significance: Comment. *Geology* **4**: 4–6.

Bailey, V. 1922. *Beaver Habits, Beaver Control, and Possibilities in Beaver Farming.* USDA Bulletin No. 1078.

Bailey, V. 1927. *Beaver Habits and Experiments in Beaver Culture.* USDA Technical Bulletin No. 21.

Bailey, V., and Bailey, F. M. 1918. *Wild Animals of Glacier National Park.* U.S. Department of the Interior, Washington, D.C.

Balch, G. C., and Jones, R. 1991. Zinc in plants, sediments, snow and ice around a galvanized electrical transmission tower in a beaver pond. *Water, Air, and Soil Pollution* **59**: 145–52.

Bardach, J. E. 1961. Transport of calcareous fragments by reef fishes. *Science* **133**: 98–9.

Barnes, W. J. 1985. Population dynamics of woody plants on a river island. *Canadian Journal of Botany* **63**: 647–55.

Barnes, W. J., and Dibble, E. 1988. The effects of beaver in riverbank forest succession. *Canadian Journal of Botany* **66**: 40–4.

Baroni, C., and Orombelli, G. 1994. Abandoned penguin rookeries as Holocene paleoclimatic indicators in Antarctica. *Geology* **22**: 23–26.

Baroni-Urbani, C., Josens, G., and Peakin, G. J. 1978. Empirical data and demographic parameters. In M. V. Brian, ed., *Production Ecology of Ants and Termites.* Cambridge University Press, Cambridge, pp. 5–44.

Basey, J. M., Jenkins, S. H., and Busher, P. E. 1988. Optimal central-place foraging by beavers: Tree-size selection in relation to defensive chemicals of quaking aspen. *Oecologia* **76**: 278–82.

Baxter, F. P., and Hole, F. D. 1967. Ant (*Formica cinerea*) pedoturbation in a prairie soil. *Soil Science Society of America Proceedings* **31**: 425–8.

Beaver build large dam at Bowman Lake. 1938. *Columbia Falls Review,* 4 April, p. 1.

Beecham, J. J., Reynolds, D. G., and Hornocker, M. G. 1983. Black bear denning activities and den characteristics in west-central Idaho. *International Conference on Bear Research and Management* **5**: 79–86.

Beier, P. 1989. Use of habitat by mountain beaver in the Sierra Nevada. *Journal of Wildlife Management* **53**: 649–54.

Beier, P., and Barrett, R. H. 1989. Beaver distribution in the Truckee River basin, California. *California Fish and Game* **75**: 233–8.

Belden, R. C., and Pelton, M. R. 1976. Wallows of the European wild hog in the mountains of east Tennessee. *Journal of the Tennessee Academy of Science* **51:** 91–3.

Berg, A. W. 1990a. Formation of Mima mounds: A seismic hypothesis. *Geology* **18:** 281–4.

Berg, A. W. 1990b. Comment and reply on "Formation of Mima mounds: A seismic hypothesis" – Reply. *Geology* **18:** 1260–1.

Berg, A. W. 1991. Comment and reply on "Formation of Mima mounds: A seismic hypothesis" – Reply. *Geology* **19:** 284–5.

Bergstrom, D. 1985. Beavers: Biologists "rediscover" a natural resource. *Forestry Research West,* USDA, pp. 1–5.

Bernard, R. T. F., and Peinke, D. 1993. Is the orientation of aardvark diggings into termitaria optional? *Die Naturwissenschaften* **80:** 422–4.

Berry, S. S. 1923. Observations on a Montana beaver canal. *Journal of Mammalogy* **4:** 92–103.

Bertram, B. C. R. 1992. *The Ostrich Communal Nesting System.* Princeton University Press, Princeton, N.J.

Best, T. L. 1972. Mound development by a pioneer population of the banner-tailed kangaroo rat, *Dipodomys spectabilis baileyi* Goldman, in eastern New Mexico. *American Midland Naturalist* **87:** 201–6.

Best, T. L., Intress, C., and Shull, K. D. 1988. Mound structure in three taxa of Mexican kangaroo rats (*Dipodomys spectabilis cratodon, D. s. zygomaticus* and *D. nelsoni*). *American Midland Naturalist* **119:** 216–20.

Beyer, W. N., Connor, E. E., and Gerould, S. 1994. Estimates of soil ingestion by wildlife. *Journal of Wildlife Management* **58:** 375–82.

Bhat, M. G., Huffaker, R. G., and Lenhart, S. M. 1993. Controlling forest damage by dispersive beaver populations: Centralized optimal management strategy. *Ecological Applications* **3:** 518–30.

Birkhead, T. R., and Harris, M. P. 1985. Ecological adaptations for breeding in the Atlantic Alcidae. In D. N. Nettleship and T. R. Birkhead, eds., *The Atlantic Alcidae.* Academic Press, London, pp. 205–31.

Black, T. A., and Montgomery, D. R. 1991. Sediment transport by burrowing mammals, Marin County, California. *Earth Surface Processes and Landforms* **16:** 163–72.

Blanchard, B. M. 1983. Grizzly bear–habitat relationships in the Yellowstone area. *International Conference on Bear Research and Management* **5:** 118–23.

Blom, P. E., Johnson, J. B., Shafii, B., and Hammel, J. 1994. Soil water movement related to distance from three *Pogonomyrmex salinus* (Hymenoptera: Formicidae) nests in south-eastern Idaho. *Journal of Arid Environments* **26:** 241–55.

Bloom, A. L. 1991. *Geomorphology.* Prentice–Hall, Englewood Cliffs, N.J.

Boesch, C., and Boesch, H. 1981. Sex differences in the use of natural hammers by wild chimpanzees: A preliminary report. *Journal of Human Evolution* **10:** 585–93.

Boesch, C., and Boesch, H. 1984. Mental map in wild chimpanzees: An analysis of hammer transports for nut cracking. *Primates* **25:** 160–70.

Bogucki, D. J., Gruendling, G. K., Allen, E. B., Adams, K. B., and Remillard, M. M. 1986. Photointerpretation of historical (1948–1985) beaver activity in the Adirondacks. *ASPRS Technical Papers, 1986 ASPRS–ACSM Fall Convention,* Anchorage, Alaska, pp. 299–308.

Bordage, G., and Filion, L. 1988. Analyse dendroécologique d'un milieu riverain fréquente par le castor (*Castor canadensis*) au Mont du Lac-des-Cygnes (Charlevoix, Québec). *Naturaliste Canadien* **115:** 117–24.

Bowman, D. M. J. S., and Panton, W. J. 1991. Sign and habitat impact of banteng (*Bos javanicus*) and pig (*Sus scrofa*), Cobourg Peninsula, northern Australia. *Australian Journal of Ecology* **16**: 15–17.

Bown, R. R. 1988. Beaver and dams: Can they coexist? In J. Emerick, S. Q. Foster, L. Hayden-Wing, J. Hodgson, J. W. Monarch, A. Smith, O. Thorne II, and J. Todd, eds., *Issues and Technology in the Management of Impacted Wildlife – Proceedings of a National Symposium.* Thorne Ecological Institute, Boulder, Colo., pp. 97–104.

Bown, T. M., and Laza, J. H. 1990. A Miocene termite nest from southern Argentina and its paleoclimatological implications. *Ichnos* **1**: 73–9.

Boyd, I. L. 1989. Spatial and temporal distribution of Antarctic fur seals (*Arctocephalus gazella*) on the breeding grounds at Bird Island, South Georgia. *Polar Biology* **10**: 179–85.

Boyer, L. F., McCall, P. L., Soster, F. M., and Whitlatch, R. B. 1990. Deep sediment mixing by burbot (*Lota lota*), Caribou Island basin, Lake Superior, USA. *Ichnos* **1**: 91–5.

Branner, J. C. 1909. Geologic work of ants in tropical America. *Geological Society of America Bulletin* **21**: 449–96.

Brazaitis, P. J. 1969. The occurrence and ingestion of gastroliths in two captive crocodilians. *Herpetologica* **25**: 63–4.

Breckenridge, W. J., and Tester, J. R. 1961. Growth, local movements and hibernation of the Manitoba toad, *Bufo hemiophrys. Ecology* **42**: 637–46.

Brian, M. V., ed., 1978. *Production Ecology of Ants and Termites.* Cambridge University Press, Cambridge.

Brodrick, H. J. 1959. *Wild Animals of Yellowstone National Park.* Yellowstone Library and Museum Association, Yellowstone Park, Wyo.

Bromley, R. G. 1978. Bioerosion of Bermuda reefs. *Palaeogeography, Palaeoclimatology, Palaeoecology* **23**: 169–97.

Bromley, R. G., and D'Alessandro, A. 1990. Comparative analysis of bioerosion in deep and shallow water, Pliocene to recent, Mediterranean Sea. *Ichnos* **1**: 43–9.

Broschart, M. R., Johnston, C. A., and Naiman, R. J. 1989. Predicting beaver colony density in boreal landscapes. *Journal of Wildlife Management* **53**: 929–34.

Brown, L. J., and Root, A. 1971. The breeding behaviour of the lesser flamingo *Phoenic-onaias minor. Ibis* **113**: 147–72.

Bubier, J. L., Moore, T. R., and Roulet, N. T. 1993. Methane emissions from wetlands in the midboreal region of northern Ontario, Canada. *Ecology* **74**: 2240–54.

Buckhouse, J. C., Skovlin, J. M., and Knight, R. W. 1981. Streambank erosion and ungulate grazing relationships. *Journal of Range Management* **34**: 339–40.

Buech, R. R., Rugg, D. J., and Miller, N. L. 1989. Temperature in beaver lodges and bank dens in a near-boreal environment. *Canadian Journal of Zoology* **67**: 1061–6.

Bullock, J. F., and Arner, D. H. 1985. Beaver damage to nonimpounded timber in Mississippi. *Southern Journal of Applied Forestry* **9**: 137–40.

Burger, J. 1993. Colony and nest site selection in lava lizards *Tropidurus* spp. in the Galapagos Islands. *Copeia* **1993**: 748–54.

Burger, J., and Gochfeld, M. 1991. Burrow site selection by black iguana (*Ctenosaura similis*) at Palo Verde, Costa Rica. *Journal of Herpetology* **25**: 430–5.

Burke, V. J., Whitfield Gibbons, J., and Green, J. L. 1994. Prolonged nesting forays by common mud turtles (*Kinosternon subrubrum*). *American Midland Naturalist* **131**: 190–5.

Burns, J. A., Flath, D. L., and Clark, T. W. 1989. On the structure and function of white-tailed prairie dog burrows. *Great Basin Naturalist* **49**: 517–24.

Burns, S. F., 1979. The northern pocket gopher (*Thomomys talpoides*): Major geomorphic agent on the alpine tundra. *Journal of the Colorado–Wyoming Academy of Science* **2:** 86.

Burrell, H. 1927. *The Platypus.* Rigby Ltd., Adelaide.

Butler, D. R. 1989. The failure of beaver dams and resulting outburst flooding: A geomorphic hazard of the southeastern Piedmont. *Geographical Bulletin* **31:** 29–38.

Butler, D. R. 1991a. Beavers as agents of biogeomorphic change: A review and suggestions for teaching exercises. *Journal of Geography* **90:** 210–17.

Butler, D. R. 1991b. The reintroduction of the beaver into the South. *Southeastern Geographer* **31:**39–43.

Butler, D. R. 1992. The grizzly bear as an erosional agent in mountainous terrain. *Zeitschrift für Geomorphologie* **36:** 179–89.

Butler, D. R. 1993. The impact of mountain goat migration on unconsolidated slopes in Glacier National Park, Montana. *Geographical Bulletin* **35:** 98–106.

Butler, D. R., and Malanson, G. P. 1985. A history of high-magnitude snow avalanches, southern Glacier National Park, Montana, USA. *Mountain Research and Development* **5:** 175–82.

Butler, D. R., and Malanson, G. P. 1989. Periglacial patterned ground, Waterton–Glacier International Peace Park, Canada and USA. *Zeitschrift für Geomorphologie* **33:** 43–57.

Butler, D. R., and Malanson, G. P. 1990. Non-equilibrium geomorphic processes and patterns on avalanche paths in the northern Rocky Mountains, USA. *Zeitschrift für Geomorphologie* **34:** 257–70.

Butler, D. R., and Malanson, G. P. 1994. Canadian landform examples – Beaver landforms. *Canadian Geographer* **38:** 76–9.

Butler, D. R., Malanson, G. P., and Kupfer, J. A. 1992. Beaver, treefall, and cutbank erosion in midwestern rivers. *Abstracts, Annual Meeting of the Association of American Geographers,* San Diego, pp. 29–30.

Butler, D. R., Malanson, G. P., and Walsh, S. J. 1991. Identification of deltaic wetlands at montane finger lakes, Montana. *Environmental Professional* **13:** 352–62.

Butler, D. R., and Schipke, K. A. 1992. The strange case of the appearing (and disappearing) lakes: The use of sequential topographic maps of Glacier National Park, Montana. *Surveying and Land Information Systems* **52:** 150–4.

Butterfield, B. R., and Key, C. H. 1986. Mapping grizzly bear habitat in Glacier National Park using a stratified Landsat classification: A pilot study. In *Proceedings – Grizzly Bear Habitat Symposium.* U.S. Forest Service General Technical Report INT-207, pp. 58–66.

Byford, J. L. 1983. *Beavers in Tennessee: Control, Utilization, and Management.* Agricultural Extension Service, University of Tennessee, Knoxville.

Bykov, A. V., and Sapanov, M. K. 1989. Significance of mammal burrowing activity in water accumulation in forest plantations on clay semidesert. *Soviet Journal of Ecology* **20:** 43–8.

Cacchione, D. A., Drake, D. E., Field, M. E., and Tate, G. B. 1987. Sea-floor gouges caused by migrating gray whales off northern California. *Continental Shelf Research* **7:** 553–60.

Cadée, G. C. 1976. Sediment reworking by *Arenicola marina* on tidal flats in the Dutch Wadden Sea. *Netherlands Journal of Sea Research* **10:** 440–60.

Cadée, G. C. 1979. Sediment reworking by the Polychaete *Heteromastus filiformis* on a tidal flat in the Dutch Wadden Sea. *Netherlands Journal of Sea Research* **13:** 441–56.

Cadée, G. C. 1989. Size-selective transport of shells by birds and its palaeoecological implications. *Palaeontology* **32:** 429–37.

Cadée, G. C. 1990. Feeding traces and bioturbation by birds on a tidal flat, Dutch Wadden Sea. *Ichnos* **1:** 23–30.

Cain, R. H. 1974. Pimple mounds: A new viewpoint. *Ecology* **55:** 178–82.

Calef, G. W., and Lortie, G. M. 1975. A mineral lick of the barren-ground caribou. *Journal of Mammalogy* **56:** 240–2.

Calkins, D. G. 1978. Feeding behavior and major prey species of the sea otter, *Enhydra lutris,* in Montague Strait, Prince William Sound, Alaska. *Fishery Bulletin* **76:** 125–31.

Campbell, H. W. 1972. Ecology or phylogenetic interpretations of crocodilian nesting habits. *Nature* **238:** 404–5.

Campbell, J. B. 1970. Hibernacula of a population of *Bufo boreas boreas* in the Colorado front range. *Herpetologica* **26:** 278–82.

Carlson, D. C., and White, E. M. 1988. Variations in surface-layer color, texture, pH, and phosphorous content across prairie dog mounds. *Soil Science Society of America Journal* **52:** 1758–61.

Carlson, K. J. 1968. The skull morphology and estivation burrows of the Permian lungfish, *Gnathorhiza serrata. Journal of Geology* **76:** 641–63.

Carlson, S. R., and Whitford, W. G. 1991. Ant mound influence on vegetation and soils in a semiarid mountain ecosystem. *American Midland Naturalist* **126:** 125–39.

Carpenter, C. C. 1982. The bullsnake as an excavator. *Journal of Herpetology* **16:** 394–401.

Carver, R. E., and Brook, G. A. 1989. Late Pleistocene paleowind directions, Atlantic Coastal Plain, USA. *Palaeogeography, Palaeoclimatology, Palaeoecology* **74:** 205–16.

Chadwick, D. H. 1983. *A Beast the Color of Winter.* Sierra Club Books, San Francisco.

Chapman, D. W. 1988. Critical review of variables used to define effects of fines in redds of large salmonids. *Transactions of the American Fisheries Society* **117:** 1–21.

Chapman, J. A., Romer, J. I., and Stark, J. 1955. Ladybird beetles and army cutworm adults as food for grizzly bears in Montana. *Ecology* **36:** 156–8.

Chesemore, D. L. 1969. Den ecology of the arctic fox in northern Alaska. *Canadian Journal of Zoology* **47:** 121–9.

Chorley, R. J., Schumm, S. A., and Sugden, D. E. 1985. *Geomorphology.* Methuen, New York.

Cincotta, R. P. 1989. Note on mound architecture of the black-tailed prairie dog. *Great Basin Naturalist* **49:** 621–3.

Clark, G. R., and Ratcliffe, B. C. 1989. Observations on the tunnel morphology of *Heterocerus brunneus* Melsheimer (Coleoptera: Heteroceridae) and its paleoecological significance. *Journal of Paleontology* **63:** 228–32.

Clark, W. L. 1993. Double mound of the harvester ant *Pogonomyrmex occidentalis* (Hymenoptera: Formicidae, Myrmicinae). *Great Basin Naturalist* **53:** 407–8.

Clayton, L. 1975. Bison trails and their geologic significance. *Geology* **3:** 498–500.

Clayton, L. 1976. Bison trails and their geologic significance: Reply. *Geology* **4:** 6–7.

Clifford, H. F., Wiley, G. M., and Casey, R. J. 1993. Macroinvertebrates of a beaver-altered boreal stream of Alberta, Canada, with special reference to the fauna on the dams. *Canadian Journal of Zoology* **71:** 1439–47.

Coetzee, B. J., Engelbrecht, A. H., Joubert, S. C. J., and Refief, P. F. 1979. Elephant impact on *Sclerocarya caffra* trees in *Acacia nigrescens* tropical plains thornveld of the Kruger National Park. *Koedoe* **22:** 39–60.

Coleman, R. L., and Dahm, C. N. 1990. Stream geomorphology: Effects on periphyton standing crop and primary production. *Journal of the North American Benthological Society* **9**: 293–302.

Colin, P. L. 1973. Burrowing behavior of the yellowhead jawfish, *Opistognathus aurifrons*. *Copeia* **1973**: 84–90.

Collias, N. E., and Collias, E. C., eds., 1976. *External Construction by Animals*. Dowden, Hutchinson and Ross, Stroudsburg, Pa.

Cook, D. O. 1971. Depressions in shallow marine sediment made by benthic fish. *Journal of Sedimentary Petrology* **41**: 577–602.

Cooper, J., and Brown, C. R. 1990. Ornithological research at the sub-Antarctic Prince Edward Islands: A review of achievements. *South African Journal of Antarctic Research* **20**: 40–57.

Costa, J. E. 1985. *Floods from Dam Failures*. USGS Open-File Report No. 85-560.

Coulombe, H. N. 1971. Behavior and population ecology of the burrowing owl, *Speotyto cunicularia*, in the Imperial Valley of California. *Condor* **73**: 162–76.

Cowan, D. P. 1991. The availability of burrows in relation to dispersal in the wild rabbit *Oryctolagus cuniculus*. *Symposium of the Zoological Society of London* **63**: 213–30.

Cowan, I. McT., and Brink, V. C. 1949. Natural game licks in the Rocky Mountain National Parks of Canada. *Journal of Mammalogy* **30**: 379–87.

Cowell, D. W. 1984. The Canadian beaver, *Castor canadensis*, as a geomorphic agent in karst terrain. *Canadian Field-Naturalist* **98**: 227–30.

Cox, G. W. 1984a. The distribution and origin of Mima mound grasslands in San Diego County, California. *Ecology* **65**: 1397–1405.

Cox, G. W. 1984b. Mounds of mystery. *Natural History* **93**: 36–45.

Cox, G. W. 1986. Mima mounds as an indicator of the chaparral–grassland boundary in San Diego County, California. *American Midland Naturalist* **116**: 64–77.

Cox, G. W. 1987. The origin of vegetation circles on stony soils of the Namib Desert near Gobabeb, South West Africa/Namibia. *Journal of Arid Environments* **13**: 237–43.

Cox, G. W. 1989. Early summer diet and food preferences of northern pocket gophers in north central Oregon. *Northwest Science* **63**: 77–82.

Cox, G. W. 1990a. Comment and reply on "Formation of Mima mounds: A seismic hypothesis" – Comment. *Geology* **18**: 1259–60.

Cox, G. W. 1990b. Form and dispersion of Mima mounds in relation to slope steepness and aspect on the Columbia Plateau. *Great Basin Naturalist* **50**: 21–31.

Cox, G. W. 1990c. Soil mining by pocket gophers along topographic gradients in a Mima moundfield. *Ecology* **71**: 837–43.

Cox, G. W., and Allen, D. W. 1987a. Soil translocation by pocket gophers in a Mima moundfield. *Oecologia* **72**: 207–10.

Cox, G. W., and Allen, D. W. 1987b. Sorted stone nets and circles of the Columbia Plateau: A hypothesis. *Northwest Science* **61**: 179–85.

Cox, G. W., and Gakahu, C. G. 1983. Mima mounds in the Kenya highlands: Significance for the Dalquest–Scheffer hypothesis. *Oecologia* **57**: 170–4.

Cox, G. W., and Gakahu, C. G. 1984. The formation of Mima mounds in the Kenya highlands: A test of the Dalquest–Scheffer hypothesis. *Journal of Mammalogy* **65**: 149–52.

Cox, G. W., and Gakahu, C. G. 1985. Mima mound microtopography and vegetation pattern in Kenyan savannas. *Journal of Tropical Ecology* **1**: 23–36.

Cox, G. W., and Gakahu, C. G. 1986. A latitudinal test of the fossorial rodent hypothesis of Mima mound origin. *Zeitschrift für Geomorphologie* **30**: 485–501.

Cox, G. W., and Gakahu, C. G. 1987. Biogeographical relationships of rhizomyid mole rats with Mima mound terrain in the Kenya highlands. *Pedobiologia* **30:** 263–75.

Cox, G. W., Gakahu, C. G., and Allen, D. W. 1987. Small-stone content of Mima mounds of the Columbia Plateau and Rocky Mountains regions: Implications for mound origin. *Great Basin Naturalist* **47:** 609–19.

Cox, G. W., Gakahu, C. G., and Waithaka, J. M. 1989. The form and small stone content of large earth mounds constructed by mole rats and termites in Kenya. *Pedobiologia* **33:** 307–14.

Cox, G. W., and Hunt, J. 1990a. Form of Mima mounds in relation to occupancy by pocket gophers. *Journal of Mammalogy* **71:** 90–4.

Cox, G. W., and Hunt, J. 1990b. Nature and origin of stone stripes on the Columbia Plateau. *Landscape Ecology* **5:** 53–64.

Cox, G. W., Lovegrove, B. G., and Siegfried, W. R. 1987. The small stone content of Mima-like mounds in the South African Cape region: Implications for mound origin. *Catena* **14:** 165–76.

Cox, G. W., Mills, J. N., and Ellis, B. A. 1992. Fire ants (Hymenoptera: Formicidae) as major agents of landscape development. *Environmental Entomology* **21:** 281–6.

Cox, G. W., and Roig, V. G. 1986. Argentinian Mima mounds occupied by ctenomyid rodents. *Journal of Mammalogy* **67:** 428–32.

Cox, G. W., and Scheffer, V. B. 1991. Pocket gophers and Mima terrain in North America. *Natural Areas Journal* **11:** 193–8.

Cox, N. J. 1989. Book review of Viles, H., ed., *Biogeomorphology. Progress in Physical Geography* **13:** 620–4.

Craighead, F. C., Jr., and Craighead, J. J. 1972. *Grizzly Bear Prehibernation and Denning Activities as Determined by Radiotracking.* Wildlife Monographs No. 32, Wildlife Society.

Crawford, H. S., Hooper, R. G., and Harlow, R. F. 1976. Woody plants selected by beavers in the Appalachian Ridge and Valley Province. USDA Forest Service Research Paper No. NE-346.

Crisp, D. T., and Carling, P. A. 1989. Observations on siting, dimensions and structure of salmonid redds. *Journal of Fish Biology* **34:** 119–34.

Crowcroft, P. 1957. *The Life of the Shrew.* Max Reinhardt, London.

Culver, D. C., and Beattie, A. J. 1983. Effects of ant mounds on soil chemistry and vegetation patterns in a Colorado montane meadow. *Ecology* **64:** 485–92.

Daborn, G. R., Amos, C. L., Brylinsky, M., Christian, H., Drapeau, G., Faas, R. W., Grant, J., Long, B., Paterson, D. M., Perillo, G. M. E., and Piccolo, M. C. 1993. An ecological cascade effect: migratory birds affect stability of intertidal sediments. *Limnology and Oceanography* **38:** 225–31.

Dalquest, W. W., and Scheffer, V. B. 1942. The origin of the Mima mounds of western Washington. *Journal of Geology* **50:** 68–84.

Dalquest, W. W., Stangl, Jr., F. B., and Kocurko, M. J. 1990. Zoogeographic implications of Holocene mammal remains from ancient beaver ponds in Oklahoma and New Mexico. *Southwestern Naturalist* **35:** 105–10.

Darlington, J. P. E. C. 1985. Lenticular soil mounds in the Kenya highlands. *Oecologia* **66:** 116–21.

Davies, A. G., and Baillie, I. C. 1988. Soil-eating by red leaf monkeys (*Presbytis rubicunda*) in Sabah, northern Borneo. *Biotropica* **20:** 252–8.

Davis, M. A., Villinski, J., Banks, K., Buckman-Fifield, J., Dicus, J., and Hofmann, S. 1991. Combined effects of fire, mound-building by pocket gophers, root loss and plant size on growth and reproduction in *Penstemon grandiflorus. American Midland Naturalist* **125:** 150–61.

Dean, W. R. J., and Milton, S. J. 1991a. Disturbances in semi-arid shrubland and arid grassland in the Karoo, South Africa: Mammal diggings as germination sites. *African Journal of Ecology* **29:** 11–16.

Dean, W. R. J., and Milton, S. J. 1991b. Patch disturbances in arid grassy dunes: Antelope, rodents and annual plants. *Journal of Arid Environments* **20:** 231–7.

Dean, W. R. J., and Siegfried, W. R. 1991. Orientation of diggings in the aardvark. *Journal of Mammalogy* **72:** 823–4.

Dean, W. R. J., and Yeaton, R. I. 1993. The effects of harvester ant *Messor capensis* nest-mounds on the physical and chemical properties of soils in the southern Karoo, South Africa. *Journal of Arid Environments* **25:** 249–60.

Del Moral, R. 1984. The impact of the Olympic marmot on subalpine vegetation structure. *American Journal of Botany* **71:** 1228–36.

De Wilde, A. W. J. 1991. Interactions in burrowing communities and their effects on the structure of marine benthic ecosystems. *Symposium of the Zoological Society of London* **63:** 107–17.

Dickinson, N. R. 1971. Aerial photographs as an aid in beaver management. *New York Fish and Game Journal* **18:** 57–61.

Dieter, C. D. 1992. Population characteristics of beavers in eastern South Dakota. *American Midland Naturalist* **128:** 191–6.

Dieter, C. D., and McCabe, T. R. 1989a. Factors influencing beaver lodge-site selection on a prairie river. *American Midland Naturalist* **122:** 408–11.

Dieter, C. D., and McCabe, T. R. 1989b. Habitat use by beaver along the Big Sioux River in eastern South Dakota. In R. E. Gresswell, B. A. Barton, and J. L. Kershner, eds., *Practical Approaches to Riparian Resource Management – An Educational Workshop.* U.S. Bureau of Land Management, Billings, Mont., pp. 135–40.

Dillon, W. P., and Zimmerman, H. B. 1970. Erosion by biological activity in two New England submarine canyons. *Journal of Sedimentary Petrology* **40:** 542–7.

Dionne, J.-C. 1985. Tidal marsh erosion by geese, St. Lawrence estuary, Québec. *Géographie Physique et Quaternaire* **39:** 99–105.

Ditmars, R. L. 1936. *Reptiles of the World.* New York: Macmillan.

Dmitriyev, P. P. 1989. Change in a soil profile due to burrowing of mammals. *Soviet Soil Science* **21:** 23–30.

Dmitriyev, P. P., Khudyakov, O. I., and Galsan, P. 1989. Micromosaic pattern of solonchak depressions in the eastern Mongolian steppes as a result of activity of burrowing mammals. *Soviet Journal of Ecology* **20:** 264–9.

Dodge, N. N. 1964. *Organ Pipe Cactus National Monument.* National Park Service Natural Handbook Series No. 6, Washington, D.C.

Dubec, L. J., Krohn, W. B., and Owen, R. B., Jr. 1990. Predicting occurrence of river otters by habitat on Mount Desert Island, Maine. *Journal of Wildlife Management* **54:** 594–9.

Dugmore, A. R. 1914. *The Romance of the Beaver.* J. B. Lippincott, Philadelphia.

Duncan, S. L. 1984. Leaving it to beaver. *Environment* **26:** 41–5.

Dyer, P. K., and Hill, G. J. E. 1990. Nearest neighbour analysis and wedge-tailed shearwater burrow patterns on Heron and Masthead Islands, Great Barrier Reef. *Australian Geographical Studies* **28:** 51–61.

Dyer, P. K., and Hill, G. J. E. 1991. A solution to the problem of determining the occupancy status of wedge-tailed shearwater *Puffinus pacificus* burrows. *Emu* **91:** 20–5.

Dyer, P. K., and Hill, G. J. E. 1992. Active breeding burrows of the wedge-tailed shearwater in the Capricorn Group, Great Barrier Reef. *Emu* **92:** 147–51.

The eager beaver – Forestry's continuing nuisance. 1986. *Georgia Forestry* **39:** 6–7.

Easter-Pilcher, A. 1990. Cache size as an index to beaver colony size in northwestern Montana. *Wildlife Society Bulletin* **18:** 110–13.

Easterbrook, D. J. 1993. *Surface Processes and Landforms.* Macmillan, New York.

Eaton, W. P. 1917. A Rocky Mountain game trail. *Harper's Magazine* **136:** 111–23.

Eberhardt, L. E., Garrott, R. A., and Hanson, W. C. 1983. Den use by arctic foxes in northern Alaska. *Journal of Mammalogy* **64:** 97–102.

Echternach, J. L., and Rose, R. K. 1987. Use of woody vegetation by beavers in southeastern Virginia. *Virginia Journal of Science* **38:** 226–32.

Edge, W. D., Marcum, C. L., and Olson-Edge, S. L. 1990. Distribution and grizzly bear, *Ursus arctos,* use of yellow sweetvetch, *Hedysarum sulphurescens,* in northwestern Montana and southeastern British Columbia. *Canadian Field-Naturalist* **104:** 435–8.

Edwards, J. K., and Guynn, D. C., Jr. 1984. Utilization of woody vegetation by beaver within the South Carolina Piedmont. Clemson University Forestry Bulletin No. 42, 1–9.

Edwards, W. M., Shipitalo, M. J., Owens, L. B., and Norton, L. D. 1990. Effect of *Lumbricus terrestris* L. burrows on hydrology of continuous no-till corn fields. *Geoderma* **46:** 73–84.

Elgmork, K. 1982. Caching behavior of brown bears (*Ursus arctos*). *Journal of Mammalogy* **63:** 607–12.

Ellison, G. T. H. 1993. Group size, burrow structure and hoarding activity of pouched mice (*Saccostomus campestris*: Cricetidae) in southern Africa. *African Journal of Ecology* **31:** 135–55.

Ellison, L. 1946. The pocket gopher in relation to soil erosion on mountain range. *Ecology* **27:** 101–14.

Elmes, G. W. 1991. Ant colonies and environmental disturbance. *Symposium of the Zoological Society of London* **63:** 15–31.

Engeman, R. M., Campbell, D. L., and Evans, J. 1991. An evaluation of 2 activity indicators for use in mountain beaver burrow systems. *Wildlife Society Bulletin* **19:** 413–16.

Ermer, E. M. 1988. Managing beaver in New York. *Conservationist* **42:** 36–9.

Evans, M. E. G. 1991. Ground beetles and the soil: Their adaptations and environmental effects. *Symposium of the Zoological Society of London* **63:** 119–32.

Ezzell, C. 1991. Hungry whales take a bite out of the beach. *Science News* **137:** 167.

Farrar, J. 1992. Musquash, grazer of the marsh. *Nebraskaland* **70:** 14–29.

Ferguson, M. W. J. 1985. Reproductive biology and embryology of the crocodilians. In C. Gans, ed., *Biology of the Reptilia,* vol. 14A. John Wiley and Sons, New York.

Ffolliott, P. F., Clary, W. P., and Larson, F. R. 1976. Observations of beaver activity in an extreme environment. *Southwestern Naturalist* **21:** 131–3.

Fiddelke, M. 1992. Beavers cause major damage in north Fulton. *Alpharetta–Roswell Revue,* January 15–31, 1992, pp. 1, 6.

Fischer, R. 1990. Biogenetic and nonbiogenetically determined morphologies of the Costa Rican Pacific coast. *Zeitschrift für Geomorphologie* **34:** 313–21.

Fisher, M. 1993. Fine-scale distributions of tropical animal mounds: A revised statistical analysis. *Journal of Tropical Ecology* **9:** 339–48.

Fitch-Snyder, H., and Lance, V. A. 1993. Behavioral observations of lithophagy in captive juvenile alligators. *Journal of Herpetology* **27:** 335–7.

Flint, R. F., and Bond, G. 1968. Pleistocene sand ridges and pans in western Rhodesia. *Geological Society of America Bulletin* **79:** 299–314.

Forbus, K., and Allen, F. 1981. *Southern Beaver Control.* Georgia Forestry Commission Research Paper No. 23, Atlanta.

Ford, T. E., and Naiman, F. J. 1988. Alteration of carbon cycling by beaver: Methane evasion rates from boreal forest streams and rivers. *Canadian Journal of Zoology* **66**: 529–33.

Formanowicz, D. R., Jr., and Ducey, P. K. 1991. Burrowing behavior and soil manipulation by a tarantula, *Rhechostica hentzi* (Girard, 1853) (Araneida: Theraphosidae). *Canadian Journal of Zoology* **69**: 2916–18.

Foster, M. S., and Schiel, D. R. 1988. Kelp communities and sea otters: Keystone species or just another brick in the wall? In G. R. VanBlaricom and J. A. Estes, eds., *The Community Ecology of Sea Otters.* Springer–Verlag, New York, pp. 92–115.

Francis, M. M., Naiman, R. J., and Melillo, J. M. 1985. Nitrogen fixation in subarctic streams influenced by beaver (*Castor canadensis*). *Hydrobiologia* **121**: 193–202.

Fraser, D., and Hristienko, H. 1981. Activity of moose and white-tailed deer at mineral springs. *Canadian Journal of Zoology* **59**: 1991–2000.

Fraser, D., and Reardon, E. 1980. Attraction of wild ungulates to mineral-rich springs in central Canada. *Holarctic Ecology* **3**: 36–40.

Friese, C. F., and Allen, M. F. 1993. The interaction of harvester ants and vesicular–arbuscular mycorrhizal fungi in a patchy semi-arid environment: The effects of mound structure on fungal dispersion and establishment. *Functional Ecology* **7**: 13–20.

Frost, K. J., Lowry, L. F., and Carroll, G. 1993. Beluga whale and spotted seal use of a coastal lagoon system in the northeastern Chukchi Sea. *Arctic* **46**: 8–16.

Frydl, P., and Stearn, W. C. 1978. Rate of bioerosion by parrotfish in Barbados reef environments. *Journal of Sedimentary Petrology* **48**: 1149–58.

Fryxell, J. M. 1992. Space use by beavers in relation to resource abundance. *Oikos* **64**: 474–8.

Fryxell, J. M., and Doucet, C. M. 1991. Provisioning time and central-place foraging in beavers. *Canadian Journal of Zoology* **69**: 1308–13.

Fryxell, J. M., and Doucet, C. M. 1993. Diet choice and the functional response of beavers. *Ecology* **74**: 1297–1306.

Fugler, S. R., Hunter, S., Newton, I. P., and Steele, W. K. 1987. Breeding biology of blue petrels *Halobaena caerulea* at the Prince Edward Islands. *Emu* **87**: 103–10.

Furness, R. W. 1991. The occurrence of burrow-nesting among birds and its influence on soil fertility and stability. *Symposium of the Zoological Society of London* **63**: 53–67.

Gakahu, C. G., and Cox, G. W. 1984. The occurrence and origin of Mima mound terrain in Kenya. *African Journal of Ecology* **22**: 31–42.

Ganskopp, D., and Vavra, M. 1987. Slope use by cattle, feral horses, deer, and bighorn sheep. *Northwest Science* **61**: 74–81.

Ganzhorn, J. U. 1987. Soil consumption of two groups of semi-free-ranging lemurs (*Lemur catta* and *Lemur fulvus*). *Ethology* **74**: 146–54.

Gardner, J. S., Smith, D. J., and Desloges, J. R. 1983. *The Dynamic Geomorphology of the Mt. Rae Area: A High Mountain Region in Southwestern Alberta.* University of Waterloo Department of Geography Publication Series No. 19, Waterloo, Ontario.

Garner, J. A., and Landers, J. L. 1981. Foods and habitat of the gopher tortoise in southwestern Georgia. *Proceedings of the Annual Conference of the Southeastern Association of Fish and Wildlife Agencies* **35**: 120–34.

Garrott, R. A., Eberhardt, L. E., and Hanson, W. C. 1983. Arctic fox den identification and characteristics in northern Alaska. *Canadian Journal of Zoology* **61**: 423–6.

Gates, C. A., and Tanner, G. W. 1988. Effects of prescribed burning on herbaceous vegetation and pocket gophers (*Geomys pinetis*) in a sandhill community. *Florida Scientist* **51**: 129–39.

Gaunt, A. S. 1965. Fossorial adaptations in the bank swallow, *Riparia riparia* (Linnaeus). *University of Kansas Science Bulletin* **46**: 99–146.

Gauthier, M., and Thomas, D. W. 1993. Nest site selection and cost of nest building by Cliff Swallows (*Hirundo pyrrhonota*). *Canadian Journal of Zoology* **71**: 1120–3.

Geist, V. 1971. *Mountain Sheep – A Study in Behavior and Evolution.* University of Chicago Press, Chicago.

Gessaman, J. A., and MacMahon, J. A. 1984. Mammals in ecosystems: Their effects on the composition and production of vegetation. *Acta Zoologica Fennica* **172**: 11–18.

Gill, D. 1972. The evolution of a discrete beaver habitat in the Mackenzie River delta, Northwest Territories. *Canadian Field-Naturalist* **86**: 233–9.

Gilmore, B. 1993. "Not recommended, but do-able." *Going to the Sun, Journal of the Glacier Mountaineering Society* **24**: 6–8.

Glozier, C., and Lee, T. 1991. Rain swamps metro Atlanta, highest amount in 113 years. *Atlanta Constitution,* June 19, 1991, pp. A1, A9.

Godfrey, G., and Crowcroft, P. 1960. *The Life of the Mole.* Museum Press, London.

Goodwin, T. M., and Marion, W. R. 1978. Aspects of the nesting ecology of American alligators (*Alligator mississippiensis*) in north-central Florida. *Herpetologica* **34**: 43–7.

Gore, J. A., and Baker, W. W. 1989. Beavers residing in caves in northern Florida. *Journal of Mammalogy* **70**: 677–8.

Gotie, R. F., and Jenks, D. L. 1984. Assessment of the use of wetlands inventory maps for determining potential beaver habitat. *New York Fish and Game Journal* **31**: 55–62.

Goudie, A. S. 1988. The geomorphological role of termites and earthworms in the tropics. In H. A. Viles, ed., *Biogeomorphology.* Basil Blackwell, New York, pp. 166–92.

Goudie, A. [S.] 1993. Human influence in geomorphology. *Geomorphology* **7**: 37–59.

Goudie, A. S., and Thomas, D. S. G. 1985. Pans in southern Africa with particular reference to South Africa and Zimbabwe. *Zeitschrift für Geomorphologie* **29**: 1–19.

Gow, C. E. 1992. Gnawing of rock outcrops by porcupines. *South African Journal of Geology* **95**: 74–5.

Grant, C. 1945. A biological explanation of the Carolina Bays. *Scientific Monthly* **61**: 443–50.

Gray, M. T. 1990. Denver's urban beavers: A gnawing problem. *Colorado Outdoors* **39**: 27–9.

Green, G. A., and Anthony, R. G. 1989. Nesting success and habitat relationships of burrowing owls in the Columbia Basin, Oregon. *Condor* **91**: 347–54.

Greer, 1970. Evolutionary and systematic significance of crocodilian nesting habits. *Nature* **227**: 523–4.

Gregory, K. J. 1988. Curriculum development in geomorphology. *Journal of Geography in Higher Education* **12**: 21–30.

Gregory, M. R., Ballance, P. F., Gibson, G. W., and Ayling, A. M. 1979. On how some rays (Elasmobranchia) excavate feeding depressions by jetting water. *Journal of Sedimentary Petrology* **49**: 1125–30.

Le Groupe Français de Géomorphologie. 1989. *Recent Advances in French Geomorphology.* Centre National de la Recherche Scientifique, Paris.

Griffiths, M. 1978. *The Biology of the Monotremes.* Academic Press, New York.

Guilcher, A. 1988. *Coral Reef Geomorphology.* John Wiley and Sons, Chichester.

Gutterman, Y. 1982. Observations on the feeding habits of the Indian crested porcupine (*Hystrix indica*) and the distribution of some hemicryptophytes and geophytes in the Negev Desert highlands. *Journal of Arid Environments* **5**: 261–8.

Gutterman, Y. 1987. Dynamics of porcupine (*Hystrix indica* Kerr) diggings: Their role in the survival and renewal of geophytes and hemicryptophytes in the Negev Desert highlands. *Israel Journal of Botany* **36**: 133–43.

Gutterman, Y., Golan, T., and Garsani, M. 1990. Porcupine diggings as a unique ecological system in a desert environment. *Oecologia* **85:** 122–127.

Gutterman, Y., and Herr, N. 1981. Influences of porcupine (*Hystrix indica*) activity on the slopes of the northern Negev Mountains – Germination and vegetation renewal in different geomorphological types and slope directions. *Oecologia* **51:** 332–4.

Hair, J. D., Hepp, G. T., Luckett, L. M., Reese, K. P., and Woodward, D. K. 1978. Beaver pond ecosystems and their relationships to multi-use natural resource management. In *Strategies for Protection and Management of Floodplain Wetlands and Other Riparian Ecosystems*. USDA General Technical Report No. WO-12, pp. 80–91.

Hall, K. J., and Williams, A. J. 1981. Animals as agents of erosion at sub-Antarctic Marion Island. *South African Journal of Antarctic Research* **10–11:** 18–24.

Hall, K. R. L., and Schaller, G. B. 1964. Tool-using behavior of the California sea otter. *Journal of Mammalogy* **45:** 287–98.

Hamer, D., and Herrero, S. 1987. Wildfire's influence on grizzly bear feeding ecology in Banff National Park, Alberta. *International Conference on Bear Research and Management* **7:** 179–86.

Hamer, D., Herrero, S., and Brady, K. 1991. Food and habitat used by grizzly bears, *Ursus arctos,* along the Continental Divide in Waterton Lakes National Park, Alberta. *Canadian Field-Naturalist* **105:** 325–9.

Hamilton, R. J., and Marchinton, R. J. 1980. Denning and related activities of black bears in the coastal plain of North Carolina. *International Conference on Bear Research and Management* **4:** 122–6.

Hammerson, G. A. 1994. Beaver (*Castor canadensis*): Ecosystem alterations, management, and monitoring. *Natural Areas Journal* **14:** 44–57.

Hansell, M. H. 1984. *Animal Architecture and Building Behaviour.* Longman, London.

Hansell, M. H. 1993. The ecological impact of animal nests and burrows. *Functional Ecology* **7:** 5–12.

Hansen, R. M., and Morris, M. J. 1968. Movement of rocks by northern pocket gopher. *Journal of Mammalogy* **49:** 391–9.

Harper, F. 1955. *The Barren Ground Caribou of Keewatin.* University of Kansas, Lawrence.

Harris, M. P., and Birkhead, T. R. 1985. Breeding ecology of the Atlantic Alcidae. In D. N. Nettleship and T. R. Birkhead, eds., *The Atlantic Alcidae*. Academic Press, London, pp. 156–204.

Harrison, C. S. 1979. The association of marine birds and feeding gray whales. *Condor* **81:** 93–5.

Hasiotis, S. T., and Mitchell, C. E. 1993. A comparison of crayfish burrow morphologies: Triassic and Holocene fossil, paleo- and neo-ichnological evidence, and the identification of their burrowing signatures. *Ichnos* **2:** 291–314.

Hasiotis, S. T., Mitchell, C. E., and Dubiel, R. F. 1993. Applications of morphologic burrow interpretations to discern continental burrow architects: Lungfish or crayfish? *Ichnos* **2:** 315–33.

Hatough-Bouran, A. 1990. The burrowing habits of desertic rodents *Jaculus jaculus* and *Gerbillus dasyurus* in the Shaumari Reserve in Jordan. *Mammalia* **54:** 341–59.

Hawkins, L. K., and Nicoletto, P. F. 1992. Kangaroo rat burrows structure the spatial organization of ground-dwelling animals in a semiarid grassland. *Journal of Arid Environments* **23:** 199–208.

Haynes, G. 1991. *Mammoths, Mastodonts, and Elephants.* Cambridge University Press, New York.

Hayward, G. D. 1989. Historical grizzly bear trends in Glacier National Park, Montana: A critique. *Wildlife Society Bulletin* **17:** 195–7.

Hazelhoff, I., Van Hoof, P., Imeson, A. C., and Kwaad, F. J. P. M. 1981. The exposure of forest soil to erosion by earthworms. *Earth Surface Processes and Landforms* **6:** 235–50.

Heard, D. C., and Williams, T. W. 1992. Distribution of wolf dens on migratory caribou ranges in the Northwest Territories, Canada. *Canadian Journal of Zoology* **70:** 1504–10.

Hebert, D., and Cowan, I. McT. 1971. Natural salt licks as a part of the ecology of the mountain goat. *Canadian Journal of Zoology* **49:** 605–10.

Hedeen, S. E. 1985. Return of the beaver, *Castor canadensis,* to the Cincinnati region. *Ohio Journal of Science* **85:** 202–3.

Heine, J. C., and Speir, T. W. 1989. Ornithogenic soils of the Cape Bird Adélie penguin rookeries, Antarctica. *Polar Biology* **10:** 89–99.

Herraro, S., McCrory, W., and Pelchat, B. 1986. Using grizzly bear habitat evaluations to locate trails and campsites in Kananaskis Provincial Park. *International Conference on Bear Research and Management* **6:** 187–93.

Heth, G. 1991. The environmental impact of subterranean mole rats (*Spalax ehrenbergi*) and their burrows. *Symposium of the Zoological Society of London* **63:** 265–80.

Heymann, E. W., and Hartmann, G. 1991. Geophagy in moustached tamarins, *Saguinus mystax* (Platyrrhini: Callitrichidae), at the Rio Blanco, Peruvian Amazonia. *Primates* **32:** 533–7.

Hibbard, E. A. 1958. Movements of beaver transplanted in North Dakota. *Journal of Wildlife Management* **22:** 209–11.

Hickman, G. C., and Brown, L. N. 1973. Mound-building behavior of the southeastern pocket gopher (*Geomys pinetis*). *Journal of Mammalogy* **54:** 786–9, 971–4.

Higgins, C. G. 1982. Grazing-step terracettes and their significance. *Zeitschrift für Geomorphologie* **26:** 459–72.

Hill, G. J. E., and Barnes, A. 1989. Census and distribution of wedge-tailed shearwater *Puffinus pacificus* burrows on Heron Island, November 1985. *Emu* **89:** 135–9.

Hines, A. H., and Loughlin, T. R. 1980. Observations of sea otters digging for clams at Monterey Harbor, California. *Fishery Bulletin* **78:** 159–63.

Hirsch, K. J. K., Stubbendieck, J., and Case, R. M. 1984. Relationships between vegetation, soils, and pocket gophers in the Nebraska Sand Hills. *Transactions of the Nebraska Academy of Sciences* **12:** 5–11.

Hobbs, H. H., Jr. 1981. The crayfishes of Georgia. *Smithsonian Contributions to Zoology* **318:** 1–549.

Hobbs, H. H., Jr., and Whiteman, M. 1991. Notes on the burrows, behavior, and color of the crayfish *Fallicambarus (F.) devastator* (Decapoda: Cambaridae). *Southwestern Naturalist* **36:** 127–35.

Hobbs, R. J., and Mooney, H. A. 1991. Effects of rainfall variability and gopher disturbance on serpentine annual grassland dynamics. *Ecology* **72:** 59–68.

Hobson, K. A. 1989. Pebbles in nests of double-crested cormorants. *Wilson Bulletin* **101:** 107–8.

Holcroft, A. C., and Herrero, S. 1984. Grizzly bear digging sites for *Hedysarum sulphurescens* roots in southwestern Alberta. *Canadian Journal of Zoology* **62:** 2571–5.

Hole, F. D. 1981. Effects of animals on soil. *Geoderma* **25:** 75–112.

Holl, S., and Bleich, V. C. 1987. Mineral lick use by mountain sheep in the San Gabriel mountains, California. *Journal of Wildlife Management* **51:** 383–5.

Hooper, J. H. D. 1958. Bat erosion as a factor in cave formation. *Nature* **182:** 1464.

Howard, J. D., Mayou, T. V., and Heard, R. W. 1977. Biogenic sedimentary structures formed by rays. *Journal of Sedimentary Petrology* **47:** 339–46.

Howard, R. J., and Larson, J. S. 1985. A stream habitat classification system for beaver. *Journal of Wildlife Management* **49**: 19–25.

Huntly, N. J. 1987. Influence of refuging consumers (pikas: *Ochotona princeps*) on subalpine meadow vegetation. *Ecology* **68**: 274–83.

Huntly, N. J., and Inouye, R. 1988. Pocket gophers in ecosystems: Patterns and mechanisms. *BioScience* **38**: 786–93.

Imeson, A. C. 1976. Some effects of burrowing animals on slope processes in the Luxembourg Ardennes, pt. 1: Excavation of animal burrows in experimental plots. *Geografiska Annaler* **58A**: 115–25.

Imeson, A. C. 1977. Splash erosion, animal activity and sediment supply in a small forested Luxembourg catchment. *Earth Surface Processes* **2**: 153–60.

Imeson, A. C., and Kwaad, F. J. P. M. 1976. Some effects of burrowing animals on slope processes in the Luxembourg Ardennes, pt. 2: The erosion of animal mounds by splash under forest. *Geografiska Annaler* **58A**: 317–28.

Interagency Grizzly Bear Committee, 1987. *Grizzly Bear Compendium.* National Wildlife Federation, Washington, D.C.

Ives, R. L. 1942. The beaver–meadow complex. *Journal of Geomorphology* **5**: 191–203.

Izawa, K. 1993. Soil-eating by *Aloutta* and *Ateles*. *International Journal of Primatology* **14**: 229–42.

James, S. W. 1991. Soil, nitrogen, phosphorus, and organic matter processing by earthworms in tallgrass prairie. *Ecology* **72**: 2101–9.

Jefferies, R. L. 1988. Pattern and process in Arctic coastal vegetation in response to foraging by lesser snow geese. In M. J. A. Werger, P. J. M. van der Aart, H. J. During, and J. T. A. Verhoeven, eds., *Plant Form and Vegetation Structure*. SPB Academic Publishing, The Hague, pp. 281–300.

Jefferies, R. L., Jensen, A., and Abraham, K. F. 1979. Vegetational development and the effect of geese on vegetation at La Pérouse Bay, Manitoba. *Canadian Journal of Botany* **57**: 1439–50.

Joanen, T., and McNease, L. L. 1989. Ecology and physiology of nesting and early development of the American alligator. *American Zoologist* **29**: 987–98.

John, R. D. 1991. Observations on soil requirements for nesting bank swallows, *Riparia riparia*. *Canadian Field-Naturalist* **105**: 251–4.

Johnson, D. L. 1989. Subsurface stone lines, stone zones, artifact manuport layers, and biomantles produced by bioturbation via pocket gophers (*Thomomys bottae*). *American Antiquity* **54**: 370–89.

Johnson, D. L. 1990. Biomantle evolution and the redistribution of earth materials and artifacts. *Soil Science* **149**: 84–102.

Johnson, D. L. 1993. Dynamic denudation evolution of tropical, subtropical and temperate landscapes with three-tiered soils: Toward a general theory of landscape evolution. *Quaternary International* **17**: 67–78.

Johnson, D. L., and Balek, C. L. 1991. The genesis of Quaternary landscapes with stonelines. *Physical Geography* **12**: 385–95.

Johnson, D. L., Watson-Stegner, D., Johnson, D. N., and Schaetzl, R. J. 1987. Proisotropic and proanisotropic processes of pedoturbation. *Soil Science* **143**: 278–92.

Johnson, K. G., and Pelton, M. R. 1983. Bedding behavior of black bears in Tennessee. *Proceedings of the Annual Conference of the Southeastern Association of Fish and Wildlife Agencies* **37**: 237–43.

Johnson, K. R., and Nelson, C. H. 1984. Side-scan sonar assessment of gray whale feeding in the Bering Sea. *Science* **225**: 1150–2.

Johnson, M. K., and Aldred, D. R. 1984. Controlling beaver in the Gulf Coastal Plain. *Proceedings of the Annual Conference, Southeastern Association of Fish and Wildlife Agencies* **38:** 189–196.

Johnson, P. L., and Billings, W. D. 1962. The alpine vegetation of the Beartooth Plateau in relation to cryopedogenic processes and patterns. *Ecological Monographs* **32:** 105–35.

Johnston, C. A., and Naiman, R. J. 1987. Boundary dynamics at the aquatic–terrestrial interface: The influence of beaver and geomorphology. *Landscape Ecology* **1:** 47–57.

Johnston, C. A., and Naiman, R. J. 1990a. Aquatic patch creation in relation to beaver population trends. *Ecology* **71:** 1617–21.

Johnston, C. A., and Naiman, R. J. 1990b. Browse selection by beaver: Effects on riparian forest composition. *Canadian Journal of Forest Research* **20:** 1036–43.

Johnston, C. A., and Naiman, R. J. 1990c. The use of a geographic information system to analyze long-term landscape alteration by beaver. *Landscape Ecology* **4:** 5–19.

Johnston, C. A., Pastor, J., and Naiman, R. J. 1993. Effects of beaver and moose on boreal forest landscapes. In R. Haines-Young, D. R. Green, and S. Cousins, eds., *Landscape Ecology and Geographic Information Systems*. Taylor and Francis, London, pp. 237–54.

Joly, Y., Frenot, Y., and Vernon, P. 1987. Environmental modifications of a subantarctic peat-bog by the wandering albatross (*Diomedea exulans*): A preliminary study. *Polar Biology* **8:** 61–72.

Jonca, E. 1972. Winter denudation of molehills in mountainous areas. *Acta Theriologica* **17:** 407–17.

Jones, R. L., and Hanson, H. C. 1985. *Mineral Licks, Geophagy, and Biogeochemistry of North American Ungulates*. Iowa State University Press, Ames.

Jones, R. S., Gutherz, E. J., Nelson, W. R., and Matlock, G. C. 1989. Burrow utilization by yellowedge grouper, *Epinephelus flavolimbatus*, in the northwestern Gulf of Mexico. *Environmental Biology of Fishes* **26:** 277–84.

Judd, S. L., Knight, R. L., and Knight, B. M. 1986. Denning of grizzly bears in the Yellowstone National Park area. *International Conference on Bear Research and Management* **6:** 111–17.

Jungerius, P. D., Van den Ancker, J. A. M., and Van Zon, H. J. M. 1989. Long term measurements of forest soil exposure and creep in Luxembourg. *Catena* **16:** 437–47.

Kaczor, S. A., and Hartnett, D. C. 1990. Gopher tortoise (*Gopherus polyphemus*) effects on soils and vegetation in a Florida sandhill community. *American Midland Naturalist* **123:** 100–11.

Kaiser, G. W., and Forbes, L. S. 1992. Climatic and oceanographic influences on island use in four burrow-nesting alcids. *Ornis Scandinavica* **23:** 1–6.

Kalisz, P. J., and Stone, E. L. 1984. Soil mixing by scarab beetles and pocket gophers in north-central Florida. *Soil Science Society of America Journal* **48:** 169–72.

Keating, K. A. 1986. Historical grizzly bear trends in Glacier National Park, Montana. *Wildlife Society Bulletin* **14:** 83–7.

Keating, K. A. 1989. Historical grizzly bear trends in Glacier National Park, Montana: A response. *Wildlife Society Bulletin* **17:** 198–201.

Kendall, K. C. 1983. Use of pine nuts by grizzly and black bears in the Yellowstone area. *International Conference on Bear Research and Management* **5:** 166–73.

Kennedy, J. P., and Brockman, H. L. 1965. Stomach stone in the American alligator, *Alligator mississippiensis* Daudin. *British Journal of Herpetology* **4:** 103–4.

King-Webster, W. A., and Kenny, J. S. 1958. Bat erosion as a factor in cave development. *Nature* **181:** 1813.

Klaus, A. D., Oliver, J. S., and Kvitek, R. G. 1990. The effects of gray whale, walrus, and ice gouging disturbance on benthic communities in the Bering Sea and Chukchi Sea, Alaska. *National Geographic Research* **6**: 470–84.

Klein, D. R., and Thing, H. 1989. Chemical elements in mineral licks and associated muskoxen feces in Jameson Land, northeast Greenland. *Canadian Journal of Zoology* **67**: 1092–5.

Klenner, W., and Kroeker, D. W. 1990. Denning behavior of black bears, *Ursus americanus*, in western Manitoba. *Canadian Field-Naturalist* **104**: 540–4.

Knapp, P. A. 1989. Natural recovery of compacted soils in semiarid Montana. *Physical Geography* **10**: 165–75.

Knight, R. R., and Mudge, M. R. 1967. Characteristics of some natural licks in the Sun River area, Montana. *Journal of Wildlife Management* **31**: 293–9.

Knopf, F. L., and Balph, D. F. 1969. Badgers plug burrows to confine prey. *Journal of Mammalogy* **50**: 635–6.

Knowles, C. J., Stoner, C. J., and Gieb, S. P. 1982. Selective use of black-tailed prairie dog towns by Mountain Plovers. *Condor* **84**: 71–4.

Kofron, C. P. 1989. Nesting ecology of the Nile crocodile (*Crocodylus niloticus*). *African Journal of Ecology* **27**: 335–41.

Kok, D., DuPreez, L. H., and Channing, A. 1989. Channel construction by the African bullfrog: Another Anuran parental care strategy. *Journal of Herpetology* **23**: 435–7.

Kolb, H. H. 1985. The burrow structure of the European rabbit (*Oryctolagus cuniculus* L.). *Journal of Zoology, London* **206**: 253–62.

Kolb, H. H. 1991a. Use of burrows and movements by wild rabbits (*Oryctolagus cuniculus*) on an area of sand dunes. *Journal of Applied Ecology* **28**: 879–91.

Kolb, H. H. 1991b. Use of burrows and movements of wild rabbits (*Oryctolagus cuniculus*) in an area of hill grazing and forestry. *Journal of Applied Ecology* **28**: 892–905.

Kolb, H. H. 1994. The use of cover and burrows by a population of rabbits (Mammalia: *Oryctolagus cuniculus*) in eastern Scotland. *Journal of Zoology, London* **233**: 9–17.

Kolenosky, G. B., and Strathearn, S. M. 1987. Winter denning of black bears in east-central Ontario. *International Conference on Bear Research and Management* **7**: 305–16.

Kondolf, G. M. 1993. Lag in stream channel adjustment to livestock exclusure, White Mountains, California. *Restoration Ecology* **1**: 226–30.

Kondolf, G. M., Cada, G. F., Sale, M. J., and Felando, T. 1991. Distribution and stability of potential salmonid spawning gravels in steep boulder-bed streams of the eastern Sierra Nevada. *Transactions of the American Fisheries Society* **120**: 177–86.

Kondolf, G. M., Cook, S. S., Maddux, H. R., and Persons, W. R. 1989. Spawning gravels of rainbow trout in Glen and Grand Canyons, Arizona. *Journal of the Arizona–Nevada Academy of Science* **23**: 19–28.

Kondolf, G. M., Sale, M. J., and Wolman, M. G. 1993. Modification of fluvial gravel size by spawning salmonids. *Water Resources Research* **29**: 2265–74.

Krylova, T. V., and Deistfel'dt, L. A. 1987. Principles of habitat distribution of mountain gopher settlements. *Soviet Journal of Ecology* **18**: 51–7.

Kushlan, J. A., and Mazzotti, F. J. 1984. Environmental effects on a coastal population of gopher tortoises. *Journal of Herpetology* **18**: 231–9.

Kvitek, R. G., Fukayama, A. K., Anderson, B. S., and Grimm, B. K. 1988. Sea otter foraging on deep-burrowing bivalves in a California coastal lagoon. *Marine Biology* **98**: 157–67.

Kvitek, R. G., and Oliver, J. S. 1986. Side-scan sonar estimates of the utilization of gray whale feeding grounds along Vancouver Island. *Continental Shelf Research* **6**: 639–54.

Kvitek, R. G., and Oliver, J. S. 1988. Sea otter foraging habits and effects on prey populations and communities in soft-bottom environments. In G. R. VanBlaricom and J. A. Estes, eds., *The Community Ecology of Sea Otters*. Springer–Verlag, New York, pp. 22–47.

Lacher, T. E., Jr., Egler, I., Alho, C. J., and Mares, M. A. 1986. Termite community composition and mound characteristics in two grassland formations in central Brazil. *Biotropica* **18**: 356–9.

Lacki, M. J., and Lancia, R. A. 1983. Changes in soil properties of forests rooted by wild boars. *Proceedings of the Annual Conference, Southeastern Association of Fish and Wildlife Agencies* **37**: 228–36.

LaCock, G. D. 1988. Effect of substrate and ambient temperatures on burrowing African penguins. *Wilson Bulletin* **100**: 131–2.

Laundré, J. W. 1989. Horizontal and vertical diameter of burrows of five small mammal species in southeastern Idaho. *Great Basin Naturalist* **49**: 646–9.

Laundré, J. W. 1990. Soil moisture patterns below mounds of harvester ants. *Journal of Range Management* **43**: 10–12.

Laundré, J. W. 1993. Effects of small mammal burrows on water infiltration in a cool desert environment. *Oecologia* **94**: 43–8.

Laundré, J. W., and Reynolds, T. D. 1993. Effects of soil structure on burrow characteristics of five small mammal species. *Great Basin Naturalist* **53**: 358–66.

Lawrence, W. H. 1952. Evidence of the age of beaver ponds. *Journal of Wildlife Management* **16**: 69–79.

Laws, R. M. 1970. Elephants as agents of habitat and landscape change in East Africa. *Oikos* **21**: 1–15.

LeCount, A. L. 1983. Denning ecology of black bears in central Arizona. *International Conference on Bear Research and Management* **5**: 71–8.

Lee, J. A. 1986. Origin of mounds under creosote bush (*Larrea tridentata*) on terraces of the Salt River, Arizona. *Journal of the Arizona–Nevada Academy of Science* **21**: 23–7.

Leidholt-Bruner, K., Hibbs, D. E., and McComb, W. C. 1992. Beaver dam locations and their effects on distribution and abundance of coho salmon fry in two coastal Oregon streams. *Northwest Science* **66**: 218–23.

Leopold, L. B., and Maddock, T., Jr. 1953. The hydraulic geometry of stream channels and some physiographic implications. USGS Professional Paper No. 252, Washington, D.C.

Lessa, E. P., and Thaeler, C. S., Jr. 1989. A reassessment of morphological specializations for digging in pocket gophers. *Journal of Mammalogy* **70**: 689–700.

Lide, R. F. 1991. *Hydrology of a Carolina Bay Located on the Upper Coastal Plain, Western South Carolina*. Unpublished Master's thesis, Department of Geography, University of Georgia.

Lindzey, F. G. 1976. Characteristics of the natal den of the badger. *Northwest Science* **50**: 178–80.

Lips, K. R. 1991. Vertebrates associated with tortoise (*Gopherus polyphemus*) burrows in four habitats in south-central Florida. *Journal of Herpetology* **25**: 477–81.

Lisle, T. E. 1989. Sediment transport and resulting deposition in spawning gravels, north coastal California. *Water Resources Research* **25**: 1303–19.

Lobeck, A. K. 1939. *Geomorphology*. McGraw–Hill, New York.

Lock, J. M. 1972. The effects of hippopotamus grazing on grasslands. *Journal of Ecology* **60**: 445–67.

Löffler, E., and Margules, C. 1980. Wombats detected from space. *Remote Sensing of Environment* **9**: 47–56.

Long, C. A., and Killingley, C. A. 1983. *The Badgers of the World.* Charles C. Thomas Publ., Springfield, Ill.

Lovegrove, B. G. 1991. Mima-like mounds (*heuweltjies*) of South Africa: The topographical, ecological and economic impact of burrowing animals. *Symposium of the Zoological Society of London* **63**: 183–98.

Lovegrove, B. G., and Painting, S. 1987. Variations in the foraging behaviour and burrow structures of the Damara molerat *Cryptomys damarensis* in the Kalahari Gemsbok National Park. *Koedoe* **30**: 149–63.

Lowry, L. F., Frost, K. J., and Burns, J. J. 1980. Feeding of bearded seals in the Bering and Chukchi Seas and trophic interactions with Pacific walruses. *Arctic* **33**: 330–42.

Luken, J. O., and Billings, W. D. 1986. Hummock-dwelling ants and the cycling of microtopography in an Alaskan peatland. *Canadian Field-Naturalist* **100**: 69–73.

Luse, D. R., and Wilds, S. 1992. A GIS approach to modifying stocking rates on rangelands affected by prairie dogs. *Geocarto International* **7**: 45–51.

Lyman, R. L. 1988. Significance for wildlife management of the late Quaternary biogeography of mountain goats (*Oreamnos americanus*) in the Pacific Northwest, USA. *Arctic and Alpine Research* **20**: 13–23.

MacCracken, J. G., Ruesk, D. W., and Hansen, R. M. 1985. Vegetation and soils of burrowing owl nest sites in Conata Basin, South Dakota. *Condor* **87**: 152–4.

McComb, W. C., Sedell, J. R., and Buchholz, T. D. 1990. Dam-site selection by beavers in an eastern Oregon basin. *Great Basin Naturalist* **50**: 273–81.

McCoy, E. D., Mushinsky, H. R., and Wilson, D. S. 1993. Pattern in the compass orientation of gopher tortoise burrows at different spatial scales. *Global Ecology and Biogeography Letters* **3**: 33–40.

MacDonald, N. F., and Hygnstrom, S. E. 1991. Little dogs of the prairie. *Nebraskaland* **69**: 24–31.

Mace, R. D., and Bissell, G. N. 1986. Grizzly bear food resources in the flood plains and avalanche chutes of the Bob Marshall Wilderness, Montana. In *Proceedings – Grizzly Bear Habitat Symposium.* U.S. Forest Service General Technical Report INT-207, pp. 78–91.

Mace, R. D., and Jonkel, C. J. 1986. Local food habits of the grizzly bear in Montana. *International Conference on Bear Research and Management* **6**: 105–10.

McNamee, T. 1990. *The Grizzly Bear.* Penguin Books, New York.

McNaughton, S. J., Ruess, R. W., and Seagle, S. W. 1988. Large mammals and process dynamics in African ecosystems. *BioScience* **38**: 794–800.

McNulty, F. 1970. *Must They Die? The Strange Case of the Prairie Dog and the Black-footed Ferret.* Doubleday, Garden City, N.Y.

Mahaney, W. C. 1986. Environmental impact in the Afroalpine and subalpine belts of Mount Kenya, east Africa. *Mountain Research and Development* **6**: 247–56.

Mahaney, W. C. 1987. Behaviour of the African buffalo on Mount Kenya. *African Journal of Ecology* **25**: 199–202.

Mahaney, W. C. 1990. *Ice on the Equator: Quaternary Geology of Mount Kenya, East Africa.* Wm. Caxton Ltd., Sister Bay, Wisc.

Mahaney, W. C., and Boyer, M. G. 1986. Appendix: Small herbivores and their influence on landform origins. *Mountain Research and Development* **6**: 256–60.

Mahaney, W. C., Hancock, R. G. V., and Inoue, M. 1993. Geochemistry and clay mineralogy of soils eaten by Japanese macaques. *Primates* **34**: 85–91.

Mahaney, W. C., and Linyuan, Z. 1991. Removal of local alpine vegetation and overgrazing in the Dalijia Mountains, northwestern China. *Mountain Research and Development* **11**: 165–7.

Mahaney, W. C., Watts, D. P., and Hancock, R. G. V. 1990. Geophagia by mountain gorillas (*Gorilla gorilla beringei*) in the Virunga Mountains, Rwanda. *Primates* **31:** 113–20.

Malanson, G. P. 1993. *Riparian Landscapes*. Cambridge University Press, Cambridge.

Malanson, G. P., and Butler, D. R. 1986. Floristic patterns on avalanche paths in the northern Rocky Mountains, USA. *Physical Geography* **7:** 231–8.

Malanson, G. P., and Butler, D. R. 1990. Woody debris, sediment, and riparian vegetation of a subalpine river, Montana, USA. *Arctic and Alpine Research* **22:** 183–94.

Malanson, G. P., and Butler, D. R. 1991. Floristic variation among gravel bars in a subalpine river. *Arctic and Alpine Research* **23:** 273–8.

Malde, H. E. 1964. Patterned ground on the western Snake River Plain, Idaho, and its possible cold-climate origin. *Geological Society of America Bulletin* **75:** 191–207.

Mandel, R. D., and Sorenson, C. J. 1982. The role of the western harvester ant (*Pogonomyrmex occidentalis*) in soil formation. *Soil Science Society of America Journal* **46:** 785–8.

Manville, A. M., II. 1987. Den selection and use by black bears in Michigan's northern lower peninsula. *International Conference on Bear Research and Management* **7:** 317–22.

Manville, R. H. 1959. The Columbian ground squirrel in northwestern Montana. *Journal of Mammalogy* **40:** 26–45.

Marston, R. A. 1994. River entrenchment in small mountain valleys of the western USA; influence of beaver, grazing and clearcut logging. *Revue de Géographie de Lyon* **69:** 11–15.

Martin, G. H. G. 1988. Mima mounds in Kenya – A case of mistaken identity. *African Journal of Ecology* **26:** 127–33.

Martin, P. 1983. Factors influencing globe huckleberry fruit production in northwestern Montana. *International Conference on Bear Research and Management* **5:** 159–65.

Martin, R. P. 1989. Notes on Louisiana gopher tortoise (*Gopherus polyphemus*) reproduction. *Herpetological Review* **20:** 36–7.

Martinez Rica, J. P., and Pardo Ara, M. P. 1990. Initial data on erosion caused by small mammals in the Central Pyrenees (Spain). *Soviet Journal of Ecology* **21:** 20–5.

Martinka, C. J. 1971. Status and management of grizzly bears in Glacier National Park, Montana. *Transactions of the Thirty-sixth North American Wildlife and Natural Resources Conference,* Washington, D.C., pp. 312–20.

Martinka, C. J. 1972. *Habitat Relationships of Grizzly Bears in Glacier National Park, Montana*. National Park Service, West Glacier, Mont.

Martinka, C. J. 1974a. Ecological role and management of *Ursus arctos* (Carnivora) in Glacier National Park, Montana. *International Conference on Bear Research and Management* **3:** 147–56.

Martinka, C. J. 1974b. Population characteristics of grizzly bears in Glacier National Park, Montana. *Journal of Mammalogy* **55:** 21–9.

Martinka, C. J. 1974c. Preserving the natural status of grizzlies in Glacier National Park. *Wildlife Society Bulletin* **2:** 13–17.

Martinka, C. J., and Kendall, K. C. 1986. Grizzly bear habitat research in Glacier National Park, Montana. In *Proceedings – Grizzly Bear Habitat Symposium*. U.S. Forest Service General Technical Report INT-207, pp. 19–23.

Mason, C. F., and Macdonald, S. M. 1986. *Otters: Ecology and Conservation*. Cambridge University Press, Cambridge.

Mattson, D. J., Blanchard, B. M., and Knight, R. R. 1991. Food habits of Yellowstone grizzly bears, 1977–1987. *Canadian Journal of Zoology* **69:** 1619–29.

Mattson, D. J., Blanchard, B. M., and Knight, R. R. 1992. Yellowstone grizzly bear mortality, human habituation, and whitebark pine seed crops. *Journal of Wildlife Management* **56:** 432–42.

Mattson, D. J., Gillin, C. M., Benson, S. A., and Knight, R. R. 1991. Bear feeding activity at alpine insect aggregation sites in the Yellowstone ecosystem. *Canadian Journal of Zoology* **69:** 2430–5.

Meadows, A. 1991. Burrows and burrowing animals: An overview. *Symposium of the Zoological Society of London* **63:** 1–13.

Meadows, P. S. 1991. The environmental impact of burrows and burrowing animals – Conclusions and a model. *Symposium of the Zoological Society of London* **63:** 327–38.

Meadows, P. S., and Meadows, A. 1991a. The geotechnical and geochemical implications of bioturbation in marine sedimentary ecosystems. *Symposium of the Zoological Society of London* **63:** 157–81.

Meadows, P. S., and Meadows, A., eds. 1991b. *The Environmental Impact of Burrowing Animals and Animal Burrows.* Zoological Society of London, Clarendon Press, Oxford.

Mech, L. D. 1966. *The Wolves of Isle Royale.* U.S. National Park Service, Washington, D.C.

Medin, D. E., and Clary, W. P. 1990. *Bird Populations in and Adjacent to a Beaver Pond Ecosystem in Idaho.* U.S. Forest Service Intermountain Research Station Research Paper No. INT-432.

Medin, D. E., and Clary, W. P. 1991. *Small Mammals of a Beaver Pond Ecosystem and Adjacent Riparian Habitat in Idaho.* U.S. Forest Service Intermountain Research Station Research Paper No. INT-445.

Medin, D.E., and Torquemada, K. E. 1988. *Beaver in Western North America: An Annotated Bibliography, 1966 to 1986.* U.S. Forest Service General Technical Report INT-242.

Medwecka-Kornas', A., and Hawro, R. 1993. Vegetation on beaver dams in the Ojców National Park (southern Poland). *Phytocoenologia* **23:** 611–18.

Melquist, W. E., and Hornocker, M. G. 1983. Ecology of river otters in west central Idaho. *Wildlife Monographs* **83:** 1–60.

Messick, J. P., and Hornocker, M. G. 1981. Ecology of the badger in southwestern Idaho. *Wildlife Monographs* **76:** 1–53.

Messier, F., and Virgl, J. A. 1992. Differential use of bank burrows and lodges by muskrats, *Ondatra zibethicus,* in a northern marsh environment. *Canadian Journal of Zoology* **70:** 1180–4.

Messier, F., Virgl, J. A., and Marinelli, L. 1990. Density-dependent habitat selection in muskrats: A test of the ideal free distribution model. *Oecologia* **84:** 380–5.

Metz, R. 1990. Tunnels formed by mole crickets (Orthoptera: Gryllotalpidae): Paleoecological implications. *Ichnos* **1:** 139–41.

Mielke, H. W. 1977. Mound building by pocket gophers (*Geomyidae*): Their impact on soils and vegetation in North America. *Journal of Biogeography* **4:** 171–80.

Miller, M. F. 1984. Bioturbation of intertidal quartz-rich sands: A modern example and its sedimentologic and paleoecologic implications. *Journal of Geology* **92:** 201–16.

Mills, E. A. 1913. *In Beaver World.* Houghton Mifflin, New York.

Mills, E. A. 1919. *The Grizzly – Our Greatest Wild Animal.* Houghton Mifflin, New York.

Milton, S. J., Dean, W. R. J., and Siegfried, W. R. 1994. Food selection by ostrich in southern Africa. *Journal of Wildlife Management* **58:** 234–48.

Minta, S. C., and Clark, T. W. 1989. Habitat suitability analysis of potential translocation sites for black-footed ferrets in northcentral Montana. In *The Prairie Dog Ecosystem: Managing for Biological Diversity*. Montana BLM Wildlife Technical Bulletin No. 2, Billings, pp. 29–45.

Mitchell, C. C., and Niering, W. A. 1993. Vegetation change in a topogenic bog following beaver flooding. *Bulletin of the Torrey Botanical Club* **120:** 136–47.

Mitchell, P. 1988. The influences of vegetation, animals and micro-organisms on soil processes. In H. A. Viles, ed., *Biogeomorphology*. Basil Blackwell, New York, pp. 43–82.

Moe, S. R. 1993. Mineral content and wildlife use of soil licks in southwestern Nepal. *Canadian Journal of Zoology* **71:** 933–6.

Moles, R. 1992. Trampling damage to vegetation and soil cover at paths within the Burren National Park, Mullach Mor, Co. Clare. *Irish Geography* **25:** 129–37.

Mollohan, C. M. 1987. Characteristics of adult female black bear daybeds in northern Arizona. *International Conference on Bear Research and Management* **7:** 145–9.

Moore, A. G. 1990. Don't mess with a badger. *Texas Parks and Wildlife* **48:** 22–7.

Moore, G. C., and Martin, E. C. 1949. *Status of Beaver in Alabama*. Alabama Department of Conservation, Montgomery.

Moorhead, D. L., Fisher, F. M., and Whitford, W. G. 1988. Cover of spring annuals on nitrogen-rich kangaroo rat mounds in a Chihuahuan Desert grassland. *American Midland Naturalist* **120:** 443–7.

Mora, J. M. 1989. Eco-behavioral aspects of two communally nesting Iguanines and the structure of their shared nesting burrows. *Herpetologica* **45:** 293–8.

Morcombe, M. K. 1968. Mammals. In G. P. Whitley, C. F. Brodie, M. K. Morcombe, and J. R. Kinghorn, eds., *Animals of the World – Australia*. Hamlyn Publ. Group Ltd., Sydney, pp. 60–97.

Moroka, N., Beck, R. F., and Pieper, R. D. 1982. Impact of burrowing activity of the banner-tail kangaroo rat on southern New Mexico desert rangelands. *Journal of Range Management* **35:** 707–10.

Müller-Schwarze, D. 1984. *The Behavior of Penguins*. State University of New York Press, Albany.

Mun, H.-T., and Whitford, W. G. 1990. Factors affecting annual plants assemblages on banner-tailed kangaroo rat mounds. *Journal of Arid Environments* **18:** 165–73.

Munn, C. A. 1994. Macaws: Winged rainbows. *National Geographic* **184:** 118–40.

Murie, O. J. 1951. *The Elk of North America*. Stackpole Company, Harrisburg, Pa.

Murison, L. D., Murie, D. J., Morin, K. R., and da Silva Curiel, J. 1984. Foraging of the gray whale along the west coast of Vancouver Island, British Columbia. In M. L. Jones, S. L. Swartz, and S. Leatherwood, eds., *The Gray Whale Eschrichtius robustus*. Academic Press, Orlando, Fla., pp. 451–63.

Mysterud, I. 1983. Characteristics of summer beds of European brown bears in Norway. *International Conference on Bear Research and Management* **5:** 208–22.

Naiman, R. J. 1988. Animal influences on ecosystem dynamics. *BioScience* **38:** 750–2.

Naiman, R. J., DeCamps, H., Pastor, J., and Johnston, C. A. 1988. The potential importance of boundaries to fluvial ecosystems. *Journal of the North American Benthological Society* **7:** 289–306.

Naiman, R. J., Johnston, C. A., and Kelley, J. C. 1988. Alteration of North American streams by beaver. *BioScience* **38:** 753–62.

Naiman, R. J., Manning, T., and Johnston, C. A. 1991. Beaver population fluctuations and tropospheric methane emissions in boreal wetlands. *Biogeochemistry* **12:** 1–15.

Naiman, R. J., and Melillo, J. M. 1984. Nitrogen budget of a subarctic stream altered by beaver (*Castor canadensis*). *Oecologia* **62:** 150–5.

Naiman, R. J., Melillo, J. M., and Hobbie, J. E. 1986. Ecosystem alteration of a boreal forest stream by beaver (*Castor canadensis*). *Ecology* **67**: 1254–69.

Naiman, R. J., Pinay, G., Johnston, C. A., and Pastor, J. 1994. Beaver influences on the long-term biogeochemical characteristics of boreal forest drainage networks. *Ecology* **75**: 905–21.

National Park Service. 1981. *Devils Tower*. Division of Publications, National Park Service, Washington, D.C.

Neal, E. [G.] 1986. *The Natural History of Badgers*. Facts on File Publications, New York.

Neal, E. G., and Roper, T. J. 1991. The environmental impact of badgers (*Meles meles*) and their setts. *Symposium of the Zoological Society of London* **63**: 89–106.

Neff, D. J. 1959. A seventy-year history of a Colorado beaver colony. *Journal of Mammalogy* **40**: 381–7.

Nelson, C. H., and Johnson, R. K. 1987. Whales and walruses as tillers of the sea floor. *Scientific American* **256**: 112–17.

Nelson, C. H., Johnson, K. R., and Barber, J. H. 1987. Gray whale and walrus feeding excavation on the Bering shelf, Alaska. *Journal of Sedimentary Petrology* **57**: 419–30.

Nerini, M. [K.] 1984. A review of gray whale feeding ecology. In M. L. Jones, S. L. Swartz, and S. Leatherwood, eds., *The Gray Whale Eschrichtius robustus*. Academic Press, Orlando, Fla., pp. 423–50.

Nerini, M. K., and Oliver, J. S. 1983. Gray whales and the structure of the Bering Sea benthos. *Oecologia* **59**: 224–5.

Nettleship, D. N. 1972. Breeding success of the Common Puffin (*Fratercula arctica*) in different habitats at Great Island, Newfoundland. *Ecological Monographs* **42**: 234–68.

Nettleship, D. N., and Birkhead, T. R., eds., 1985. *The Atlantic Alcidae: The Evolution, Distribution and Biology of the Auks Inhabiting the Atlantic Ocean and Adjacent Water Areas*. Academic Press, London.

Nir, D. 1983. *Man, a Geomorphological Agent: An Introduction to Anthropogenic Geomorphology*. Keter, Jerusalem.

Novak, M. 1987. Beaver. In M. Novak, J. A. Baker, M. E. Obbard, and B. Malloch, eds., *Wild Furbearer Management and Conservation in North America*. Ontario Ministry of Natural Resources, Toronto, pp. 282–312.

Nummi, P. 1989. Simulated effects of the beaver on vegetation, invertebrates and ducks. *Annales Zoologici Fennici* **26**: 43–52.

Oates, J. F. 1978. Water-plant and soil consumption by guereza monkeys (*Colobus guereza*): A relationship with minerals and toxins in the diet? *Biotropica* **10**: 241–53.

Obst, B. S., and Hunt, G. L., Jr. 1990. Marine birds feed at gray whale mud plumes in the Bering Sea. *Auk* **107**: 678–88.

Ojany, F. F. 1968. The mound topography of the Thika and Athi plains of Kenya: A problem of origin. *Erdkunde* **22**: 269–75.

Oliver, J. S., Slattery, P. N., O'Connor, E. F., and Lowry, L. F. 1983a. Walrus, *Odobenus rosmarus,* feeding in the Bering Sea: A benthic perspective. *Fishery Bulletin* **81**: 501–12.

Oliver, J. S., Slattery, P. N., Silberstein, M. A., and O'Connor, E. F. 1983b. A comparison of gray whale, *Eschrichtius robustus,* feeding in the Bering Sea and Baja California. *Fishery Bulletin* **81**: 513–22.

Oliver, J. S., Slattery, P. N., Silberstein, M. A., and O'Connor, E. F. 1984. Gray whale feeding on dense ampeliscid amphipod communities near Bamfield, British Columbia. *Canadian Journal of Zoology* **62**: 41–9.

Orr, R. T. 1977. *The Little-Known Pika*. Macmillan, New York.

Osmundson, C. L., and Buskirk, S. W. 1993. Size of food caches as a predictor of beaver colony size. *Wildlife Society Bulletin* **21**: 64–9.

Ottaway, E. M., Carling, P. A., Clarke, A., and Reader, N. A. 1981. Observations on the structure of the brown trout, *Salmo trutta* Linnaeus, redds. *Journal of Fish Biology* **19**: 593–607.

Ouellet, J.-P., Heard, D. C., and Boutin, S. 1993. Range impacts following the introduction of caribou on Southampton Island, Northwest Territories, Canada. *Arctic and Alpine Research* **25**: 136–41.

Owen-Smith, R. N. 1988. *Megaherbivores: The Influence of Very Large Body Size on Ecology.* Cambridge University Press, Cambridge.

Pack, A. N. 1928. Camera hunting on the Continental Divide, or, "getting your goat." *Nature Magazine* **11**: 88–94.

Parker, B. S., Myers, K., and Caskey, R. L. 1976. An attempt at rabbit control by warren ripping in semi-arid western New South Wales. *Journal of Applied Ecology* **13**: 353–67.

Parker, M., Wood, F. J., Jr., Smith, B. H., and Elder, R. G. 1985. Erosional downcutting in lower order riparian ecosystems: Have historical changes been caused by removal of beaver? In *Riparian Ecosystems and Their Management: Reconciling Conflicting Uses.* U.S. Forest Service General Technical Report RM-120, pp. 35–8.

Parsons, G. R., and Brown, M. K. 1978. An assessment of aerial photograph interpretation for recognizing potential beaver colony sites. *New York Fish and Game Journal* **25**: 175–7.

Pastor, J., Naiman, R. J., Dewey, B., and McInnes, P. 1988. Moose, microbes, and the boreal forest. *BioScience* **38**: 770–7.

Peaker, M. 1969. Active acquisition of stomach stones in the American alligator, *Alligator mississippiensis* Daudin. *British Journal of Herpetology* **4**: 103–4.

Pearson, A. M. 1975. *The Northern Interior Grizzly Bear Ursus arctos L.* Canadian Wildlife Service Report Series No. 34, Ottawa.

Peart, D. R. 1989. Species interactions in a successional grassland. III. Effects of canopy gaps, gopher mounds and grazing on colonization. *Journal of Ecology* **77**: 267–89.

Pedevillano, C., and Wright, R. G. 1987. The influence of visitors on mountain goat activities in Glacier National Park, Montana. *Biological Conservation* **39**: 1–11.

Perez, F. L. 1992a. The ecological impact of cattle on caulescent Andean rosettes in a high Venezuelan Paramo. *Mountain Research and Development* **12**: 29–46.

Perez, F. L. 1992b. Processes of turf exfoliation (*Rasenabschälung*) in the high Venezuelan Andes. *Zeitschrift für Geomorphologie* **36**: 81–106.

Perez, F. L. 1993. Alpine turf destruction by cattle in the high equatorial Andes. *Mountain Research and Development* **13**: 107–10.

Petal, J. 1978. The role of ants in ecosystems. In M. V. Brian, ed., *Production Ecology of Ants and Termites.* Cambridge University Press, Cambridge, pp. 293–325.

Phillips, J. D. 1991. The human role in earth surface systems: Some theoretical considerations. *Geographical Analysis* **23**: 316–31.

Pinder, A. W., Storey, K. B., and Ultsch, G. R. 1992. Estivation and hibernation. In M. E. Feder and W. W. Burggren, eds., *Environmental Physiology of the Amphibians.* University of Chicago Press, Chicago, pp. 250–74.

Pinkowski, B. 1983. Foraging behavior of beavers (*Castor canadensis*) in North Dakota. *Journal of Mammalogy* **64**: 312–14.

Platt, W. J. 1975. The colonization and formation of equilibrium plant species associations on badger disturbances in a tall-grass prairie. *Ecological Monographs* **45**: 285–305.

Polis, G. A., Myers, C., and Quinlan, M. 1986. Burrowing biology and spatial distribution of desert scorpions. *Journal of Arid Environments* **10**: 137–46.

Pomeroy, D. E. 1977. The distribution and abundance of large termite mounds in Uganda. *Journal of Applied Ecology* **14**: 465–75.

Pomeroy, D. E. 1978. The abundance of large termite mounds in Uganda in relation to their environment. *Journal of Applied Ecology* **15**: 51–63.

Powell, K. L., Robel, R. J., Kemp, K. E., and Nellis, M. D. 1994. Aboveground counts of black-tailed prairie dogs: Temporal nature and relationship to burrow entrance density. *Journal of Wildlife Management* **58**: 361–6.

Prestrud, P. 1992. Denning and home-range characteristics of breeding arctic foxes in Svalbard. *Canadian Journal of Zoology* **70**: 1276–83.

Price, L. W. 1971. Geomorphic effect of the Arctic ground squirrel in an alpine environment. *Geografiska Annaler* **53A**: 100–6.

Price, W. A. 1968. Carolina Bays. In R. W. Fairbridge, ed., *The Encyclopedia of Geomorphology*. Reinhold Book Corp., New York, pp. 102–8.

Pringle, C. M., Naiman, R. J., Bretschko, G., Karr, J. R., Oswood, M. W., Webster, J. R., Welcomme, R. L., and Winterbourn, M. J. 1988. Patch dynamics in lotic systems: The stream as a mosaic. *Journal of the North American Benthological Society* **7**: 503–24.

Propper, C. R., Jones, R. E., Rand, M. S., and Austing, H. 1991. Nesting behavior of the lizard *Anolis carolinensis*. *Journal of Herpetology* **25**: 484–6.

Pullan, R. A. 1979. Termite hills in Africa: Their characteristics and evolution. *Catena* **6**: 267–91.

Pullen, T. M., Jr. 1975. Observations on construction activities of beaver in east-central Alabama. *Journal of the Alabama Academy of Science* **46**: 14–19.

Rahm, D. A. 1962. The terracette problem. *Northwest Science* **36**: 65–80.

Rains, B. 1987. Holocene alluvial sediments and a radiocarbon-dated relict beaver dam, Whitemud Creek, Edmonton, Alberta. *Canadian Geographer* **31**: 272–7.

Rand, A. S., and Dugan, B. 1983. Structure of complex iguana nests. *Copeia* **1983**: 705–11.

Reading, R. P., Grensten, J. J., Beissinger, S. R., and Clark, T. W. 1989. Attributes of black-tailed prairie dog colonies, associated species, and management implications. In *The Prairie Dog Ecosystem: Managing for Biological Diversity*. Montana BLM Wildlife Technical Bulletin No. 2, Billings, pp. 13–27.

Rebertus, A. J. 1986. Bogs as beaver habitat in north-central Minnesota. *American Midland Naturalist* **116**: 240–5.

Reichelt, A. C. 1991. Environmental effects of meiofaunal burrowing. *Symposium of the Zoological Society of London* **63**: 33–52.

Reichman, O. J., and Jarvis, J. U. M. 1989. The influence of three sympatric species of fossorial mole-rats (Bathyergidae) on vegetation. *Journal of Mammalogy* **70**: 763–71.

Reichman, O. J., Benedix, J. H., Jr., and Seastedt, T. R. 1993. Distinct animal-generated edge effects in a tallgrass prairie community. *Ecology* **74**: 1281–5.

Reid, D. G., Herrero, S. M., and Code, T. E. 1988. River otters as agents of water loss from beaver ponds. *Journal of Mammalogy* **69**: 100–7.

Remillard, M. M., Gruendling, G. K., and Bogucki, D. J. 1987. Disturbance by beaver (*Castor canadensis* Kuhl) and increased landscape heterogeneity. In M. G. Turner, ed., *Landscape Heterogeneity and Disturbance*. Springer–Verlag, New York, pp. 103–22.

Resh, V. H., Brown, A. V., Covich, A. P., Gurtz, M. E., Li, H. W., Minshall, G. W., Reice, S. R., Sheldon, A. L., Wallace, J. B., and Wissmar, R. C. 1988. The role of disturbance in stream ecology. *Journal of the North American Benthological Society* **7**: 433–55.

Reynolds, H. V., Curatolo, J. A., and Quimby, R. 1976. Denning ecology of grizzly bears in northeastern Alaska. *International Conference on Bear Research and Management* **3**: 403–9.

Reynolds, P. S. 1993. Size, shape, and surface area of beaver, *Castor canadensis,* a semi-aquatic mammal. *Canadian Journal of Zoology* **71**: 876–82.

Rhoads, B. L., and Thorn, C. E. 1993. Geomorphology as science: The role of theory. *Geomorphology* **6**: 287–307.

Rice, D. W., and Wolman, A. A. 1971. *The Life History and Ecology of the Gray Whale (Eschrichtius robustus)*. Special Publication No. 3, American Society of Mammalogists.

Rice, R. J. 1977. *Fundamentals of Geomorphology*. Longman Scientific, New York.

Rich, T. 1984. Monitoring burrowing owl populations: Implications of burrow re-use. *Wildlife Society Bulletin* **12**: 178–80.

Rich, T. 1986. Habitat and nest-site selection by burrowing owls in the sagebrush steppe of Idaho. *Journal of Wildlife Management* **50**: 548–55.

Riedman, M. L., and Estes, J. A. 1988. A review of the history, distribution and foraging ecology of sea otters. In G. R. VanBlaricom and J. A. Estes, eds., *The Community Ecology of Sea Otters*. Springer–Verlag, New York, pp. 4–21.

Ritchie, M. E., and Belovsky, G. E. 1990. Sociality of Columbian ground squirrels in relation to their seasonal energy intake. *Oecologia* **83**: 495–503.

Ritter, D. F. 1986. *Process Geomorphology*. William C. Brown, Dubuque, Iowa.

Robel, R. J., and Fox, L. B. 1993. Comparison of aerial and ground survey techniques to determine beaver colony densities in Kansas. *Southwestern Naturalist* **38**: 357–61.

Robel, R. J., Fox, L. B., and Kemp, K. E. 1993. Relationship between habitat suitability index values and ground counts of beaver colonies in Kansas. *Wildlife Society Bulletin* **21**: 415–21.

Robinson, C. H., Pearce, T. G., and Ineson, P. 1991. Burrowing and soil consumption by earthworms in limed and unlimed soils from *Picea sitchensis* plantations. *Pedobiologia* **35**: 360–7.

Roe, F. G. 1970. *The North American Buffalo*. University of Toronto Press, Toronto, 2d ed.

Rohdenburg, H. 1989. *Landscape Ecology – Geomorphology*. Catena Verlag, Cremlingen-Destedt.

Roper, T. J. 1992. Badger *Meles meles* setts – Architecture, internal environment and function. *Mammal Review* **22**: 43–53.

Ross, B. A., Tester, J. R., and Breckenridge, W. J. 1968. Ecology of Mima-type mounds in northwestern Minnesota. *Ecology* **49**: 172–7.

Ross, P. I., Hornbeck, G. E., and Horejsi, B. L. 1988. Late denning black bears killed by grizzly bear. *Journal of Mammalogy* **69**: 818–20.

Rostagno, C. M., and del Valle, H. F. 1988. Mounds associated with shrubs in aridic soils of northeastern Patagonia: Characteristics and probable genesis. *Catena* **15**: 347–59.

Ruedemann, R., and Schoonmaker, W. J. 1938. Beaver-dams as geologic agents. *Science* **88**: 523–5.

Rumyantsev, V. Y. 1989. An experiment in mapping the distribution of the steppe marmot using aerial photographs. *Mapping Sciences and Remote Sensing* **26**: 294–9.

Rupp, R. S. 1955. Beaver–trout relationships in the headwaters of Sunkhaze Stream, Maine. *Transactions, American Fisheries Society* **84**: 75–85.

Rutherford, W. H. 1953. Effects of a summer flash flood upon a beaver population. *Journal of Mammalogy* **34**: 261–2.

Rutin, J. 1992. Geomorphic activity of rabbits on a coastal sand dune, De Blink dunes, the Netherlands. *Earth Surface Processes and Landforms* **17**: 85–94.

Rützler, K. 1975. The role of burrowing sponges in bioerosion. *Oecologia* **19**: 203–16.

Ryden, H. 1989. *Lily Pond – Four Years with a Family of Beavers.* William Morrow and Co., New York.

Saltz, D., and Alkon, P. U. 1992. Observations on den shifting in Indian crested porcupines in the Negev (Israel). *Mammalia* **56**: 665–7.

Sandgren, P., and Fredskild, B. 1991. Magnetic measurements recording late Holocene man-induced erosion in S. Greenland. *Boreas* **20**: 315–31.

San Jose, J. J., Montes, R., Stansly, P. A., and Bentley, B. L. 1989. Environmental factors related to the occurrence of mound-building nasute termites in Trachypogon savannas of the Orinoco Llanos. *Biotropica* **21**: 353–8.

Saucier, R. T. 1991. Comment and reply on "Formation of Mima mounds: A seismic hypothesis" – Comment. *Geology* **19**: 284.

Schalles, J. F., Sharitz, R. R., Gibbons, J. W., Leversee, G. J., and Knox, J. N. 1989. *Carolina Bays of the Savannah River Plant.* Savannah River Plant National Environmental Research Park Program SRO-NERP-18, Aiken, S.C.

Scheid, M. 1987. A beaver's tale. *Habitat: Journal of the Maine Audubon Society* **4**: 28–9.

Scheffer, V. B. 1958. Do fossorial rodents originate Mima-type microrelief? *American Midland Naturalist* **59**: 505–10.

Schipke, K. A., and Butler, D. R. 1991. The use of dendrogeomorphic techniques to date a beaver-dam outburst flood in Oglethorpe County, Georgia. *Geographical Bulletin* **33**: 80–6.

Schoen, J. W., Beier, L. R., Lentfer, J. W., and Johnson, L. J. 1987. Denning ecology of brown bears on Admiralty and Chichagof Islands. *International Conference on Bear Research and Management* **7**: 293–304.

Schoen, J. W., Lentfer, J. W., and Beier, L. 1986. Differential distribution of brown bears on Admiralty Island, southeast Alaska: A preliminary assessment. *International Conference on Bear Research and Management* **6**: 1–5.

Schrader, S., and Joschko, M. 1991. A method for studying the morphology of earthworm burrows and their function in respect to water movement. *Pedobiologia* **35**: 185–90.

Schwartz, C. C., Miller, C. D., and Franzmann, A. W. 1987. Denning ecology of three black bear populations in Alaska. *International Conference on Bear Research and Management* **7**: 281–91.

Scott, H. W. 1951. The geological work of the mound-building ants in western United States. *Journal of Geology* **59**: 173–5.

Scotter, G. W., and Scotter, E. 1989. Beaver, *Castor canadensis,* mortality caused by felled trees in Alberta. *Canadian Field-Naturalist* **103**: 400–1.

Seely, J. A., Zegers, G. P., and Asquith, A. 1989. Use of digger bee burrows by the tree lizard (*Urosauraus ornatus*) for winter retreats. *Herpetological Review* **20**: 6–7.

Selby, M. J. 1985. *Earth's Changing Surface.* Clarendon Press, Oxford.

Ser$veen, M. C., and Klaver, R. 1983. Grizzly bear dens and denning activity in the Mission and Rattlesnake Mountains, Montana. *International Conference on Bear Research and Management* **5**: 201–7.

Shachak, M., Brand, S., and Gutterman, Y. 1991. Porcupine disturbances and vegetation pattern along a resource gradient in a desert. *Oecologia* **88**: 141–7.

Sheets, R. G., Linder, R. L., and Dahlgren, R. B. 1971. Burrow systems of prairie dogs in South Dakota. *Journal of Mammalogy* **52**: 451–2.

Shepherd, V. 1986. Beavers – Loved or dammed? *Virginia Wildlife* **46**: 22–6.

Shipes, D. A., Fendley, T. T., and Hill, H. S. 1979. Woody vegetation as food items for South Carolina Coastal Plain beaver. *Proceedings of the Annual Conference, Southeastern Association of Fish and Wildlife Agencies* **33**: 202–11.

Simoons, F. J. 1974. Contemporary research themes in the cultural geography of domesticated animals. *Geographical Review* **64**: 557–76.

Singer, F. J. 1975. *Behavior of Mountain Goats, Elk, and Other Wildlife in Relation to U.S. Highway 2, Glacier National Park.* Federal Highway Administration, Denver.

Singer, F. J. 1976. *Seasonal Concentration of Grizzly Bears and Notes on Other Predators in the North Fork of the Flathead River.* Glacier National Park Research Note No. 2, West Glacier, Mont.

Singer, F. J. 1978a. Behavior of mountain goats in relation to U.S. Highway 2, Glacier National Park, Montana. *Journal of Wildlife Management* **42**: 591–7.

Singer, F. J. 1978b. Seasonal concentrations of grizzly bears, North Fork of the Flathead River, Montana. *Canadian Field-Naturalist* **92**: 283–6.

Singer, F. J. 1986. The mountain goats of U.S. 2. *Western Wildlands* **12**: 30–2.

Singer, F. J., and Doherty, J. L. 1985a. Managing mountain goats at a highway crossing. *Wildlife Society Bulletin* **13**: 469–77.

Singer, F. J., and Doherty, J. L. 1985b. Movements and habitat use in an unhunted population of mountain goats, *Oreamnos americanus*. *Canadian Field-Naturalist* **99**: 205–17.

Sinitsyn, M. G., and Rusanov, A. V. 1990. European beaver impact on small rivers' valley and channel relief in the Vetluga–Unzha woodlands. *Geomorfologiya* **1990**: 85–91. (In Russian, summary in English.)

Skinner, C., Skinner, P., and Harris, S. 1991. The past history and recent decline of badgers *Meles meles* in Essex: An analysis of some of the contributory factors. *Mammal Review* **21**: 67–80.

Skinner, Q. D., Smith, M. A., Dodd, J. L., and Rodgers, J. D. 1988. Reversing desertification of riparian zones along cold desert streams. In E. E. Whitehead, C. F. Hutchinson, B. N. Timmermann, and R. G. Varady, eds., *Arid Lands Today and Tomorrow.* Westview Press, Boulder, Colo., pp. 1407–14.

Skipworth, J. P. 1974. Ingestion of grit by bighorn sheep. *Journal of Wildlife Management* **38**: 880–3.

Slaymaker, O. 1993. The sediment budget of the Lillooet River basin, British Columbia. *Physical Geography* **14**: 304–20.

Slough, B. G., and Sadleir, R. M. F. S. 1977. A land capability classification system for beaver (*Castor canadensis* Kuhl). *Canadian Journal of Zoology* **55**: 1324–35.

Smith, C. A. S., Smits, C. M. M., and Slough, B. G. 1992. Landform selection and soil modifications associated with arctic fox (*Alopex lagopus*) den sites in Yukon Territory, Canada. *Arctic and Alpine Research* **24**: 329–36.

Smith, D. J., and Gardner, J. S. 1985. Geomorphic effects of ground squirrels in the Mount Rae area, Canadian Rocky Mountains. *Arctic and Alpine Research* **17**: 205–10.

Smith, D. W., and Peterson, R. O. 1991. Behavior of beaver in lakes with varying water levels in northern Minnesota. *Environmental Management* **15**: 395–401.

Smith, L. L., and Doughty, R. W. 1984. *The Amazing Armadillo.* University of Texas Press, Austin.

Smith, M. E., Driscoll, C. T., Wyskowski, B. J., Brooks, C. M., and Cosentini, C. C. 1991. Modification of stream ecosystem structure and function by beaver (*Castor canadensis*) in the Adirondack Mountains, New York. *Canadian Journal of Zoology* **69**: 55–61.

Smith, R. I. L. 1988. Destruction of Antarctic terrestrial ecosystems by a rapidly increasing fur seal population. *Biological Conservation* **45**: 55–72.

Smits, C. M. M., Smith, C. A. S., and Slough, G. B. 1988. Physical characteristics of arctic fox (*Alopex lagopus*) dens in northern Yukon Territory, Canada. *Arctic* **41**: 12–16.

Snyder, L. 1993. Puffin cavity nests. *Wildbird* 7: 76.

Sokol, O. 1971. Lithophagy and geophagy in reptiles. *Journal of Herpetology* 5: 67–71.

Sowls, L. K. 1984. *The Peccaries.* University of Arizona Press, Tucson.

Sparks, D. W., and Andersen, D. C. 1988. The relationship between habitat quality and mound building by a fossorial rodent, *Geomys bursarius. Journal of Mammalogy* 69: 583–7.

Spencer, T. 1988. Coastal biogeomorphology. In H. A. Viles, ed., *Biogeomorphology.* Basil Blackwell, New York, pp. 255–318.

Splettstoesser, J. F. 1985. Note on rock striations caused by penguin feet, Falkland Islands. *Arctic and Alpine Research* 17: 107–11.

Stanford, J. A., and Ward, J. V. 1993. An ecosystem perspective of alluvial rivers: Connectivity and the hyporheic corridor. *Journal of the North American Benthological Society* 12: 48–60.

Stock, J. D., and Schlosser, I. J. 1991. Short-term effects of a catastrophic beaver dam collapse on a stream fish community. *Environmental Biology of Fishes* 31: 123–9.

Stockstad, D. S., Morris, M. S., and Lory, E. C. 1953. Chemical characteristics of natural licks used by big game animals in western Montana. *Transactions of the North American Wildlife Conference* 18: 247–57.

Stokes, D. L., and Boersma, P. D. 1991. Effects of substrate on the distribution of Magellanic penguin (*Spheniscus magellanicus*) burrows. *Auk* 108: 923–33.

Stone, E. L. 1993. Soil burrowing and mixing by a crayfish. *Soil Science Society of America Journal* 57: 1096–9.

Sudd, J. H., and Franks, N. R. 1987. *The Behavioural Ecology of Ants.* Chapman and Hall, New York.

Summerfield, M. A. 1991. *Global Geomorphology.* Longman Scientific, New York.

Suzuki, T., ed., 1989. Recent trend of geomorphology in Japan. *Transactions, Japanese Geomorphological Union* 10A: 1–180.

Svendsen, G. E. 1976. Structure and location of burrows of yellow-bellied marmot. *Southwestern Naturalist* 20: 487–94.

Swanson, F. J., Krantz, T. K., Caine, N., and Woodmansee, R. G. 1988. Landform effects on ecosystem patterns and processes. *BioScience* 38: 92–8.

Swihart, R. K., and Picone, P. M. 1994. Damage to apple trees associated with woodchuck burrows in orchards. *Journal of Wildlife Management* 58: 357–60.

Tadzhiyev, U., and Odinoshoyev, A. 1987. Influence of marmots on soil cover of the eastern Pamirs. *Soviet Soil Science* 2: 22–30.

Tankersley, N. G., and Gasaway, W. C. 1983. Mineral lick use by moose in Alaska. *Canadian Journal of Zoology* 61: 2242–9.

Taylor, M. 1985. The beavers are back. *Wildlife in North Carolina* 49: 22–7.

Teipner, C. L., Garton, E. O., and Nelson, L., Jr. 1983. *Pocket Gophers in Forest Ecosystems.* U.S. Forest Service General Technical Report No. INT-154.

Terres, J. K. 1980. *The Audubon Society Encyclopedia of North American Birds.* Alfred A. Knopf, New York.

Thomas, D. S. G. 1988. The biogeomorphology of arid and semi-arid environments. In H. A. Viles, ed., *Biogeomorphology.* Basil Blackwell, New York, pp. 193–221.

Thomsen, L. 1971. Behavior and ecology of burrowing owls on the Oakland Municipal Airport. *Condor* 73: 177–92.

Thorn, C. E. 1978. A preliminary assessment of the geomorphic role of pocket gophers in the alpine zone of the Colorado Front Range. *Geografiska Annaler* 60A: 181–7.

Thorn, C. E. 1982. Gopher disturbance: Its variability by Braun–Blanquet vegetation units in the Niwot Ridge alpine tundra zone, Colorado Front Range, USA. *Arctic and Alpine Research* 14: 45–51.

Thornbury, W. D., 1969. *Principles of Geomorphology.* John Wiley and Sons, New York.

Thorne, D. H., and Andersen, D. C. 1990. Long-term soil-disturbance pattern by a pocket gopher, *Geomys bursarius. Journal of Mammalogy* **71:** 84–9.

Thornes, J. B., ed., 1990. *Vegetation and Erosion – Processes and Environments.* John Wiley and Sons, Chichester.

Thouless, C. R., Fanshawe, J. H., and Bertram, B. C. R. 1989. Egyptian vultures *Neophron percnopterus* and ostrich *Struthio camelus* eggs: The origins of stone-throwing behaviour. *Ibis* **131:** 9–15.

Thurow, R. F., and King, J. G. 1994. Attributes of Yellowstone cutthroat trout redds in a tributary of the Snake River, Idaho. *Transactions of the American Fisheries Society* **123:** 37–50.

Tietje, W. D., and Ruff, R. L. 1980. Denning behavior of black bears in boreal forest of Alberta. *Journal of Wildlife Management* **44:** 858–70.

Tisdell, C. A. 1982. *Wild Pigs: Environmental Pest or Economic Resource?* Pergamon Press, Sydney.

Toland, B. 1991. Spotted skunk use of a gopher tortoise burrow for breeding. *Florida Scientist* **54:** 10–12.

Tollner, E. W., Calvert, G. V., and Langdale, G. 1990. Animal trampling effects on soil physical properties of two southeastern U.S. ultisols. *Agriculture, Ecosystems and Environment* **33:** 75–87.

Townsend, J. E. 1953. Beaver ecology in western Montana with special reference to movements. *Journal of Mammalogy* **34:** 459–79.

Trimble, S. W. 1988. The impact of organisms on overall erosion rates within catchments in temperate regions. In H. A. Viles, ed., *Biogeomorphology.* Basil Blackwell, New York, pp. 83–142.

Trimble, S. W. 1990. Geomorphic effects of vegetation cover and management: Some time and space considerations in prediction of erosion and sediment yield. In J. B. Thornes, ed., *Vegetation and Erosion.* John Wiley and Sons, Chichester, pp. 55–65.

Troy, S., and Elgar, M. A. 1991. Brush-turkey incubation mounds: Mate attraction in a promiscuous mating system. *Trends in Ecology and Evolution* **6:** 202–3.

Turcek, F. J. 1963. The role of animals in baring and soil erosion on karst-lands. *Acta Zoologica Hungaricae* **11:** 203–15.

Twichell, D. C., Grimes, C. B., Jones, R. S., and Able, K. W. 1985. The role of erosion by fish in shaping topography around Hudson Submarine Canyon. *Journal of Sedimentary Petrology* **55:** 712–19.

Twidale, C. R., 1975. *Geomorphology.* Thomas Nelson, Ltd., West Melbourne.

Ursic, S. J., and Esher, R. J. 1988. Influence of small mammals on stormflow responses of pine-covered catchments. *Water Resources Bulletin* **24:** 133–9.

Ustinov, S. K. 1976. The brown bear on Baikal: A few features of vital activity. *International Conference on Bear Research and Management* **3:** 325–6.

VanBlaricom, G. R. 1988. Effects of foraging by sea otters on mussel-dominated intertidal communities. In G. R. VanBlaricom and J. A. Estes, eds., *The Community Ecology of Sea Otters.* Springer–Verlag, New York, pp. 48–91.

Van den Berge, E. P., and Gross, M. R. 1984. Female size and nest depth in coho salmon (*Oncorhynchus kisutch*). *Canadian Journal of Fisheries and Aquatic Sciences* **41:** 204–6.

Vander Wall, S. B. 1990. *Food Hoarding in Animals.* University of Chicago Press, Chicago.

Van Lawick-Goodall, J. 1968. Tool-using bird: The Egyptian vulture. *National Geographic* **133:** 630–41.

Van Wormer, J. 1972. *The World of the Moose.* J. B. Lippincott, Philadelphia.

Veblen, T. T., Mermoz, M., Martin, C., and Kitzberger, T. 1992. Ecological impacts of introduced animals in Nahuel Huapi National Park, Argentina. *Conservation Biology* **6:** 71–83.

Verbeek, N. A. M., and Boasson, R. 1984. Local alteration of alpine calcicolous vegetation by birds: Do the birds create hummocks? *Arctic and Alpine Research* **16:** 337–41.

Vermeer, D. E., and Frate, D. A. 1975. Geophagy in a Mississippi county. *Annals of the Association of American Geographers* **65:** 414–24.

Viles, H. A. 1988a. Introduction. In H. A. Viles, ed., *Biogeomorphology.* Basil Blackwell, New York, pp. 1–8.

Viles, H. A. 1988b. Organisms and karst geomorphology. In H. A. Viles, ed., *Biogeomorphology.* Basil Blackwell, New York, pp. 319–50.

Viles, H. A. 1988c. Perspectives. In H. A. Viles, ed., *Biogeomorphology.* Basil Blackwell, New York, pp. 351–5.

Viles, H. A., ed., 1988d. *Biogeomorphology.* Basil Blackwell, New York.

Viles, H. A. 1990. "The agency of organic beings": A selective review of recent work in biogeomorphology. In J. B. Thornes, ed., *Vegetation and Erosion – Processes and Environments.* John Wiley and Sons, Chichester, pp. 5–24.

Vitek, J. D. 1978. Morphology and pattern of earth mounds in south-central Colorado. *Arctic and Alpine Research* **10:** 701–14.

Von Frisch, K. 1983. *Animal Architecture.* Van Nostrand Reinhold, New York.

Voslamber, B., and Veen, A. W. L. 1985. Digging by badgers and rabbits on some wooded slopes in Belgium. *Earth Surface Processes and Landforms* **10:** 799–82.

Vroom, G. W., Herrero, S., and Ogilvie, R. T. 1980. The ecology of winter den sites of grizzly bears in Banff National Park, Alberta. *International Conference on Bear Research and Management* **4:** 321–30.

Wainscott, V. J., Bartley, C., and Kangas, P. 1990. Effect of muskrat mounds on microbial density on plant litter. *American Midland Naturalist* **123:** 399–401.

Warren, E. R. 1905. Some interesting beaver dams in Colorado. *Proceedings of the Washington Academy of Sciences* **6:** 429–37.

Warren, E. R. 1927. *The Beaver – Its Works and Its Ways.* American Society of Mammalogists, Baltimore.

Warren, E. R. 1932. The abandonment and reoccupation of pond sites by beavers. *Journal of Mammalogy* **13:** 343–6.

Washburn, A. L. 1980. *Geocryology.* John Wiley and Sons, New York.

Wathen, W. G., Johnson, K. G., and Pelton, M. R. 1986. Characteristics of black bear dens in the southern Appalachian region. *International Conference on Bear Research and Management* **6:** 119–27.

Weir, J. S. 1969. Chemical properties and occurrence on Kalahari sand of salt licks created by elephants. *Journal of Zoology* **158:** 293–310.

Weir, J. S. 1972. Spatial distribution of elephants in an African National Park in relation to environmental sodium. *Oikos* **23:** 1–13.

Weitkamp, LA., Wissmar, R. C., Simenstad, C. A., Fresh, K. L., and Odell, J. G. 1992. Gray whale foraging on ghost shrimp (*Callianassa californiensis*) in littoral sand flats of Puget Sound, USA. *Canadian Journal of Zoology* **70:** 2275–80.

Weltz, N., Wood, M. K., and Parker, E. E. 1989. Flash grazing and trampling effects on infiltration rates and sediment yield on a selected New Mexico range site. *Journal of Arid Environments* **16:** 95–100.

Werner, T. J. 1977. Grizzly bear dens in the Border Grizzly Area. In *Annual Research Summary, Glacier National Park.* National Park Service, West Glacier, Mont., pp. 60–1.

Wheeler, A. 1991. *Crocodilians as Biogeomorphic Agents.* Unpublished paper, Geography 816, University of Georgia, Athens.

Whicker, A. D., and Detling, J. K. 1988a. Ecological consequences of prairie dog disturbances. *BioScience* **38:** 778–84.

Whicker, A. D., and Detling, J. K. 1988b. Modification of vegetation structure and ecosystem processes by North American grassland mammals. In M. J. A. Werger, P. J. M. van der Aart, H. J. During, and J. T. A. Verhoeven, eds., *Plant Form and Vegetation Structure.* SPB Academic Publishing, The Hague, pp. 301–16.

Whitford, W. G., Ludwig, J. A., and Noble, J. C. 1992. The importance of subterranean termites in semi-arid ecosystems in south-eastern Australia. *Journal of Arid Environments* **22:** 87–91.

Wigley, T. B., and Garner, M. E. 1986. Landowner-reported beaver damage in the Arkansas delta. *Proceedings of the Annual Conference, Southeastern Association of Fish and Wildlife Agencies* **40:** 476–82.

Wigley, T. B., and Garner, M. E. 1987. Landowner perceptions of beaver damage and control in Arkansas. *Proceedings of the Third Eastern Wildlife Damage Control Conference*, Gulf Shores, Ala., pp. 34–41.

Wiig, Ø., Gjertz, I., Griffiths, D., and Lydersen, C. 1993. Diving patterns of an Atlantic walrus *Odobenus rosmarus rosmarus* near Svalbard. *Polar Biology* **13:** 71–2.

Wilde, S. A., Youngberg, C. T., and Hovind, J. H. 1950. Changes in composition of ground water, soil fertility, and forest growth produced by the construction and removal of beaver dams. *Journal of Wildlife Management* **14:** 123–8.

Wilkinson, T. 1993. *Glacier Park Wildlife.* NorthWord Press, Minocqua, Wisc.

Williams, L. R., and Cameron, G. N. 1990. Dynamics of burrows of Attwater's pocket gopher (*Geomys attwateri*). *Journal of Mammalogy* **71:** 433–8.

Williams, R. B. G. 1988. The biogeomorphology of periglacial environments. In H. A. Viles, ed., *Biogeomorphology.* Basil Blackwell, New York, pp. 222–52.

Willing, B., and Sramek, R. 1989. Urban beaver damage and control in Dallas–Fort Worth, Texas. In *Ninth Great Plains Wildlife Damage Control Workshop Proceedings.* U.S. Forest Service General Technical Report No. RM-171, pp. 77–80.

Willis, C. K., Skinner, J. D., and Robertson, H. G. 1992. Abundance of ants and termites in the False Karoo and their importance in the diet of the aardvark *Orycteropus afer. African Journal of Ecology* **30:** 322–34.

Wilsson, L. 1971. Observations and experiments on the ethology of the European beaver (*Castor fiber* L.). *Swedish Wildlife* **8:** 160–5, 168, 182–203, 254–60.

Winkle, P. L., Hubert, W. A., and Rahel, F. J. 1990. Relations between brook trout standing stocks and habitat features in beaver ponds in southeastern Wyoming. *North American Journal of Fisheries Management* **10:** 72–9.

Witz, B. W., Wilson, D. S., and Palmer, M. D. 1991. Distribution of *Gopherus polyphemus* and its vertebrate symbionts in three burrow categories. *American Midland Naturalist* **126:** 152–8.

Woo, M-K., and Waddington, J. M. 1990. Effects of beaver dams on subarctic wetland hydrology. *Arctic* **43:** 223–30.

Wood, J. C., Wood, M. K., and Trombie, J. M. 1987. Important factors influencing water infiltration and sediment production on arid lands in New Mexico. *Journal of Arid Environments* **12:** 111–18.

Woodell, S. R. J., and King, T. J. 1991. The influence of mound-building ants on British lowland vegetation. In C. R. Huxley and D. F. Cutler, eds., *Ant–Plant Interactions.* Oxford University Press, Oxford, pp. 521–35.

Yair, A., and Rutin, J. 1981. Some aspects of the regional variation in the amount of available sediment produced by isopods and porcupines, northern Negev, Israel. *Earth Surface Processes and Landforms* **6**: 221–34.

Yavitt, J. B., Angell, L. L., Fahey, T. J., Cirmo, C. P., and Driscoll, C. T. 1992. Methane fluxes, concentrations, and production in two Adirondack beaver impoundments. *Limnology and Oceanography* **37**: 1057–66.

Yavitt, J. B., Lang, G. E., and Sexstone, A. J. 1990. Methane fluxes in wetland and forest soils, beaver ponds, and low-order streams of a temperate forest ecosystem. *Journal of Geophysical Research* **95**: 22,463–74.

Yeaton, R. I. 1988. Porcupines, fires and the dynamics of the tree layer of the *Burkea africana* savanna. *Journal of Ecology* **76**: 1017–29.

Yensen, E., Luscher, M. P., and Boyden, S. 1991. Structure of burrows used by the Idaho ground squirrel, *Spermophilus brunneus*. *Northwest Science* **65**: 93–100.

Young, A., and Saunders, I. 1986. Rates of surface processes and denudation. In A. D. Abrahams, ed., *Hillslope Processes*. Allen and Unwin, Winchester, Mass., pp. 3–27.

Young, P. J. 1990. Structure, location and availability of hibernacula of Columbian ground squirrels (*Spermophilus columbianus*). *American Midland Naturalist* **123**: 357–64.

Zager, P., Jonkel, C., and Habeck, J. 1983. Logging and wildfire influence on grizzly bear habitat in northwestern Montana. *International Conference on Bear Research and Management* **5**: 124–32.

Zimmerman, J. W. 1990. Burrow characteristics of the nine-banded armadillo, *Dasypus novemcinctus*. *Southwestern Naturalist* **35**: 226–7.

Zurowski, W. 1992. Building activity of beavers. *Acta Theriologica* **37**: 403–11.

Index